A Volume in The Laboratory Animal Pocket Reference Series

The Laboratory
SMALL
RUMINANT

The Laboratory Animal Pocket Reference Series

Editor-in-Chief
Mark A. Suckow, D.V.M.
Laboratory Animal Program
Purdue University
West Lafayette, IN

Advisory Board

B. Taylor Bennett, D.V.M., Ph.D.
Biologic Resources Laboratory
University of Illinois at Chicago
Chicago, IL

John Harkness, D.V.M., M.S., M.Ed.
College of Veterinary Medicine
Mississippi State University
Mississippi State, MS

Terrie Cunliffe-Beamer, D.V.M., M.S.
The Jackson Laboratory
Bar Harbor, ME

Roger P. Maickel, Ph.D.
Laboratory Animal Program
Purdue University
West Lafayette, IN

Published and Forthcoming Titles

The Laboratory Rabbit

The Laboratory Non-Human Primates

The Laboratory Mouse

The Laboratory Guinea Pig

The Laboratory Rat

The Laboratory Hamster and Gerbil

The Laboratory Cat

The Laboratory Small Ruminant

A Volume in The Laboratory Animal Pocket Reference Series

The Laboratory
SMALL
RUMINANT

Matthew J. Allen, Vet.M.B., Ph.D.

Assistant Professor
Department of Orthopedic Surgery
SUNY-Health Science Center
Syracuse, New York

Gary L. Borkowski, D.V.M., M.S.

Diplomate ACLAM
Associate Director
Laboratory Animal Resources
The Monsanto Company
St. Louis, Missouri

Editor-in-Chief
Mark A. Suckow, D.V.M.

CRC Press
Boca Raton   London   New York   Washington, D.C.

CRC Press
Taylor & Francis Group
6000 Broken Sound Parkway NW, Suite 300
Boca Raton, FL 33487-2742

© 1999 by Taylor & Francis Group, LLC
CRC Press is an imprint of Taylor & Francis Group

No claim to original U.S. Government works

10 9 8 7 6 5 4 3
International Standard Book Number-13: 978-0-8493-2568-7 (Softcover)

Library of Congress Cataloging-in-Publication Data

Catalog record is available from the Library of Congress

Visit the Taylor & Francis Web site at
http://www.taylorandfrancis.com

and the CRC Press Web site at
http://www.crcpress.com

dedication

MJA: This book is dedicated to my parents, Kate and Graham Allen, my wife, Clare, and our favorite sheep farmer, James Millett, for their support and encouragement throughout my career.

GLB: I dedicate this book to Dr. Paul Eness, Dr. Arlo Ledet, Dr. William Hoefle, Ms. Gail Wolz, and Dr. Gerald Van Hoosier.

preface

The use of laboratory animals, including small ruminants such as sheep and goats, continues to be an important part of biomedical research. In many instances, individuals performing such research have with broad responsibilities, including animal facility management, animal husbandry, regulatory compliance, and performance of technical procedures directly related to research projects. This handbook was written to provide a quick reference source for investigators, technicians, and animal caretakers charged with the care and/or use of small ruminants in a laboratory setting. It should be particularly valuable to those at small institutions or facilities lacking a centralized resource unit, and to those individuals who need to utilize small ruminants in research or education projects but lack prior experience.

This handbook is organized into six chapters: Important Biological Features (Chapter 1), Husbandry (Chapter 2), Management (Chapter 3), Veterinary Care (Chapter 4), Experimental Methodology (Chapter 5), and Resources (Chapter 6). The information presented in this handbook will allow new or inexperienced personnel to develop the skills required to perform experimental procedures on sheep and goats. References are provided for techniques or procedures that are beyond the scope of this handbook.

This handbook should be viewed as a basic reference source and not as an exhaustive review of the care and use of small ruminants. The last chapter in this book provides lists of sources and suppliers of additional information, animals, feed, sanitation supplies, cages, and research and veterinary supplies.

These lists are not exhaustive, and do not imply endorsement of listed suppliers over suppliers not listed. Rather, these lists are starting points for users to be able to develop their own lists of preferred vendors of such items.

A final point to be considered is that all individuals performing procedures described in this handbook should be properly trained. The humane care and use of small ruminants is improved by initial and continuing education of personnel. Such training will facilitate the overall success of programs using small ruminants in research, education or testing programs.

the authors

Matthew J. Allen holds a dual appointment as Assistant Professor of Orthopedic Surgery and Clinical Veterinarian at the State University of New York (SUNY) Health Science Center at Syracuse, NY. Dr. Allen graduated with a Bachelor of Veterinary Medicine (Vet.M.B.) degree from the University of Cambridge (U.K.) in 1991, and completed Ph.D. training in comparative orthopedics at Addenbrooke's Hospital in Cambridge. After a one-year fellowship in experimental orthopedics at the School of Veterinary Medicine at Purdue University, he moved to SUNY-Health Science Center at Syracuse in 1996. Dr. Allen's primary research interests are in bone biology and comparative orthopedics, including the development and refinement of animal models for total joint replacement. Much of this work has involved the use of large animal models, particularly dogs and sheep.

Gary L. Borkowski is Associate Director of Laboratory Animal Resources at the Monsanto Company, St. Louis, MO. Dr. Borkowski received his Doctor of Veterinary Medicine (D.V.M.) degree from Iowa State University in 1987 and received a Master of Science degree in Laboratory Animal Medicine from Pennsylvania State University in 1989. He is a Diplomate of the American College of Laboratory Animal Medicine (ACLAM), and has held elected and appointed positions in the American Association for Laboratory Animal Science (AALAS), ACLAM, and the American Society of Laboratory Animal Practitioners (ASLAP). Dr. Borkowski has more than 40 publications and presentations in the field of laboratory animal science on topics including anesthesia, euthanasia, aseptic surgery, animals in space research, computer resources and utilization, and institutional animal care and use committees (IACUCs).

acknowledgment

The authors wish to acknowledge the generous contributions of Dr. Paul Eness, Ms. Susan DeRienzo. Mr. Tim Allen, Mr. Gary Keller, Dr. Peter Jackson, Mr. Murray Corke, and Dr. Simon Turner.

contents

important biological features

introduction

Small ruminants have been used in research for many years and, in contrast with species such as the dog, cat, and non-human primate, their use appears to be increasing. Published figures from the United States Department of Agriculture (USDA) show that 72,331 sheep were used in biomedical research, teaching, and testing in 1996.[1] Annual statistics for the use of goats are not reported separately by the USDA, as goats are classified as members of the broad category of "farm animals". Since small ruminants used in agricultural research are also exempt from reporting by the USDA, the statistics for sheep and goats clearly under-represent the real numbers of small ruminants used in research in the United States.

origins of the sheep and goat

Sheep and goats were among the first animals to be domesticated. Archeological evidence suggests that sheep were being raised for wool production as long ago as 4000 B.C., while goat remains have been dated to between 6000–7000 B.C.[2] Although the precise origins of domestic sheep remain controversial, it is generally agreed that they originated from the wild Mouflon, which it still found in both Western Asia and Europe.[3] The predecessor of the modern goat is the Bezoar from Asia Minor and the Middle East.[2] Taxonomically, sheep and goats can be classified as:

Order: Artiodactyla (even-toed ungulates)

Family: Bovidae (common ruminants; cattle, sheep, goats, and antelopes)

Species: *Ovis aries* (domestic sheep)
          Capra hircus (domestic goat)

Selection for sheep (on the basis of meat, wool, milk, and skin) has been so successful that there are now over 1 billion sheep of at least 300 breeds distributed across the globe. The greatest numbers are in Australia, Africa, Europe, and China.[3,4]

Selection for goats (on the basis of milk, meat, mohair, and skin) has led to a worldwide population of at least 609.5 million, with the greatest numbers in the developing tropical regions of the world.[4] In temperate countries, goats are raised primarily for hair, hides, meat, and milk, while in arid, semitropical, and mountainous regions, they have the added advantage of filling an important ecological niche, being able to graze land on which sheep and cattle simply could not survive. Goats are also more heat tolerant, and have lower metabolic and water requirements than sheep.[5]

breeds

A complete description of the more than 300 sheep and 45 goat breeds that have been recorded is beyond the scope of this book, but a number of breeds are popular in agricultural and biomedical research. The choice of a specific breed of sheep or goat for a study is usually based upon factors such as the type of housing available at an institution, the size of animal required for a study, and previous reports of the suitability of certain breeds. Sheep and goats are extremely flexible in this regard, since they can be housed under conditions as diverse as state-of-the-art indoor vivaria, indoor-outdoor barns, and commercial-style outdoor ranches. Depending on the breed and age of the animals that are chosen, sheep and goats can be obtained at body weights that range from as low as 20 kg (mature dwarf goat) to as high as 200 kg (mature Suffolk ram).

Popular sheep breeds include the Suffolk and Dorset (from England), Finn (from Finland), and Merino (from Australia), while popular goat breeds include the Pygmy (originally from West Africa), Nubian and Spanish dairy goats, Saanen and Toggenburg (both from Switzerland). Figures 1 and 2 illustrate some of these breeds.

behavior

Sheep are for the most part docile, non-aggressive, gregarious creatures.[6] From a husbandry perspective, the most important behavioral characteristics of sheep are their tendency to flock, their short flight distance, and their dependence on strong visual cues.[7]

The short flight distance in sheep means that it is possible to get quite close to a sheep before it tries to run away. When challenged, sheep will run toward one another, forming a group; although forcing the animals together in this way can facilitate their capture, in a laboratory setting this can lead to injuries, particularly if the animals charge toward a narrow opening (e.g., a doorway). Gentle, hands-off control of sheep is much more effective, and the simple act of moving or waving an arm in a certain direction will cause the sheep to move in the opposite direction. Once one animal has been enticed to move in the desired direction, the remainder of the flock will generally follow.

Visual cues are extremely important to sheep. They dislike changes in lighting conditions and shadows. Sheep prefer to move from dark toward light, and this can be exploited when trying to load animals onto vehicles or along a raceway. Isolation is extremely stressful to sheep, and they should be in direct visual range of other sheep at all times.

Aggressive behavior is usually restricted to rams, although as in most species, a ewe with lambs at foot can present a threat to intruders. In rams, aggression is typically seen as head

Fig. 1.   Common sheep breeds: **(A)** Suffolk; **(B)** Horned Dorset.

Fig. 2. Common goat breeds: **(A)** Saanen; **(B)** Nubian dairy goat.

clashing; care should be taken when handling rams, as they are capable of inflicting serious injury on personnel. Ewes can also exhibit aggression, although this is usually restricted to a less forceful form of head butting, particularly around the lambing season.

Goats have the most obvious social hierarchy of farm animals. The dominant buck (the alpha buck) is aggressive during the breeding season, but for the remainder of the year he may be submissive to the leading doe (the queen). Once the social hierarchy has been established, it is usually extremely stable.[7] The introduction of new animals into the herd can, however, upset the equilibrium and fights may well develop. Aggressive behavior is primarily seen in bucks. In contrast with rams, bucks rear up before clashing heads.

Goats are extremely adventurous and inquisitive, being browsers rather than grazers. This has significant implications for their housing, since they will lick and chew at painted surfaces (leading to the risk of lead poisoning),[8] lean and push against fences, and jump over any obstruction that is not high enough (a minimum height of 4' 6" is recommended for fences). They are fastidious eaters, refusing food or water that has become soiled with feces or urine, a fact that must be taken into account when pens are designed. Hay should be fed in racks to allow the goats to browse, water troughs should be elevated off the ground, and concentrates should be supplied well away from areas where the animals defecate.

It is impossible to drive goats. When challenged by an intruder, they stomp one forefoot and may emit a high-pitched sneezing sound. If the intruder advances, the animals initially form up in a line, then they scatter, relying on their agility over uneven surfaces to evade capture.

anatomical and physiological features

Important and unique anatomic and physiologic features of the sheep and goat include:

Dentition

- The dental formula of both sheep and goats is 2 (I 0/4: C 0/0; P 3/3; M 3/3).

- The upper incisors are absent, their place being taken by a fibrous pad (the dental pad).

- The teeth do not erupt continuously and are prone to attrition. Uneven wear, particularly of the cheek teeth (molars and premolars) will lead to malocclusion.

- Dental diseases, including malocclusion and dental caries, are significant causes of morbidity in commercial sheep and goat enterprises.[8,9]

Skeleton

- The axial skeleton of sheep and goats consists of 7 cervical vertebrae, 12 to 14 thoracic vertebrae, 6 to 7 lumbar vertebrae, 4 sacral vertebrae and 16 to 18 coccygeal vertebrae.

- Both sheep and goats have an upright stance, with the distal limb reduced to two main digits plus two rudimentary digits on the caudal surface of the metacarpus and metatarsus.

- The only significant difference between the skeleton of sheep and goats is in the development of skull bones, which are much more developed in the male goat. This adaptation arises from their normal aggressive behavior (see previous).

Cutaneous Structures

- The most obvious cutaneous structure in sheep and goats is the **hair coat**.

- In sheep, natural selection for wool production has led to a wide variety of fleece types, with tremendous variation between breeds in terms of the cut (total fleece production per year), length, and mean fiber diameter of the final product.

- Most sheep have either white or black faces and cream/off-white fleeces, although colored fleeces are seen in Welsh Mountain sheep (black all over) and rare breeds such as the Jacob (multicolored fleece).

- Goats have also been selected for hair and fiber production, leading to the development of the Angora (in Turkey, South Africa, and the United States), the Cashmere (in Afghanistan), and the Don (in Russia) for mohair and cashmere wool production.

- The hair coat of goats can be almost any color, ranging from white to black and including mixtures of colors (as patches or stripes, or as blended shades).

- Goats (but not sheep) have natural **scent glands** located around the horn buds in both sexes, as well as in the tail of the male. The secretions from these glands can become offensive in animals that are housed indoors.

- The goat tail is short, turns up, and is bare on its ventral surface.

- Horns may be present in both sheep and goats. In goats, possession of horns is linked to the polled gene, which is an autosomal dominant.[10] However, the polled gene is also linked to a recessive gene associated with abnormal sexual development.

- Removal of horn buds from young goats requires care since the skull is thin in this area, and excessive heating may produce cerebral damage and/or meningitis.[11]

- Goats of both sexes may have **beards** and **wattles**.

- Other differences between sheep and goats include the absence of **lacrimal pits** (below the eyes) in goats. Goats also lack **interdigital glands** in their hindfeet and **wax glands** in the groin region.

Gastrointestinal System

- Both sheep and goats are true ruminants, possessing a four-part stomach divided into the reticulum, rumen, omasum (collectively known as the forestomach) and abomasum (the "true" stomach).

- The anatomical and functional organization of the ruminant digestive apparatus has evolved as a tremendously efficient system for converting a variety of food sources into volatile fatty acids, the fundamental fuel of ruminant metabolism.

- In young lambs and kids, milk bypasses the reticulorumen by means of the reticular groove. The reticular groove operates until the animal is approximately 8 weeks old and ready to be weaned; from this time on, lambs and kids function as true ruminants, with ingesta passing from the esophagus into the reticulorumen.

- Within the rumen, food is digested by protozoa and bacteria and mixed by cyclical contractions of the wall of the reticulorumen.

- Eructation, the voiding of volatile gases from the rumen, is normal in both sheep and goats.

- Rumination, the process by which boluses of partially digested food are regurgitated and subjected to a second cycle of mastication, is unique to the ruminants.

- The feces of sheep and goat is normally produced as clumps of pellets.

Urogenital System

- The walls of the cervix are thrown into a series of interlocking folds, making it extremely difficult to pass even a catheter through the cervix.

- The urethra of young male sheep is relatively narrow and is predisposed to obstruction at the level of the urethral process (see Urolithiasis, Chapter 4).

- The placenta is epitheliochorial, with the exchange of nutrients and waste products occurring at regions of direct contact between placental villi and maternal capillary plexuses.[10,12] These areas of contact are known as cotyledons.

normative values

Typical values for basic biological parameters (Table 1), clinical chemistry (Table 2), hematology (Table 3), cerebrospinal fluid (Table 4), and respiratory and cardiovascular performance (Table 5) follow.

Note: Normative values can vary significantly between clinical laboratories, as well as between animals of different breeds, ages and sexes. The following tables are intended to provide only broad guidelines.

TABLE 1. BASIC BIOLOGICAL PARAMETERS OF SHEEP AND GOATS

Parameter	Typical Values		Reference(s)
	Sheep	Goat	
Diploid chromosome number	54	60	10,12,13
Life span (years)	10–15	10–15	10,12,13
Adult body weight (kg)	20–160	10–70	10,12,13
Body temperature (°C)	38.0–39.5	38.0–39.5	8,10,12
Metabolic rate (Kcal/kg$^{0.75}$/day)	47–62	65.3	14,15
Dry matter intake (g/kg/day)	15–60	24–60	8,16
Water intake (ml/kg/day)	197	188	17
Fecal output (kg/day)	1–3	0.5–3	18
Urine volume (ml/kg/day)	10–40	10–40	19
Urine specific gravity	1.015–1.045	1.001–1.050	19
Urine pH	7.5–8.5	7.2–8.5	8,19

TABLE 2. CLINICAL CHEMISTRY VALUES OF SHEEP AND GOATS

Parameter	Typical Values		Reference(s)
	Sheep	Goat	
Total protein (g/l)	60–80	55.8–86.4	20
Albumin (g/l)	35–45	55.8–86.4	20
Globulin (g/l)	35–57	27–39	8
Alkaline phosphatase (ALP) IU/L	50–300	41.0–1195	20
Lactate dehydrogenase (LDH) IU/L	238–440	217.0–586.0	8,21
γ-Glutamyl transferase (γ-GT) IU/L	40–94	2.6–67.7	20
Aspartate aminotransferase (AST) IU/L	60–280	43–142	20
Creatine phosphokinase (CPK) IU/L	100–547	143.0–678.0	20
Sorbitol dehydrogenase (SDH) IU/L	18–77	2–57	20
Urea (mmol/l)	2.9–7.1	3.4–11.5	20
Creatinine (mmol/l)	106–168	48.0–136.0	20
Glucose (mmol/l)	2.8–4.4	2.6–4.1	20
Sodium (mmol/l)	139–152	141.0–157.0	20,22
Chloride (mmol/l)	95–103	102.0–113.0	20,22
Potassium (mmol/l)	3.9–5.4	3.50–7.08	20
Calcium (mmol/l)	2.9–3.2	2.15–2.81	20
Phosphorus (mEq/L)	1.6–2.4	1.7–4.3	8,20
Total bilirubin (µmol/l)	0.7–3.8	0–0.1	23
Cholesterol (mmol/l)	1.1–2.3	2.07–3.36	8,24

TABLE 3. SELECTED NORMAL VALUES FOR CEREBROSPINAL FLUID IN SHEEP AND GOATS

Parameter	Typical Values		Reference(s)
	Sheep	Goat	
Total protein (mg/dl)	8–70	12	25
Total white blood cells ($\times 10^3$/ml)	0–5	0–9	26
Glucose (mmol/l)	1.7	3.1	27,28
pH	7.3–7.4	7.3–7.4	8,25
Specific gravity	1.004–1.008	1.004–1.008	8

TABLE 4. VALUES FOR CARDIOVASCULAR AND RESPIRATORY FUNCTION IN SHEEP AND GOATS

Parameter	Typical Values		Reference(s)
	Sheep	Goat	
Respiratory rate (breaths/min)	10–20	10–30	18
Heart rate (beats/min)	60–120	70–135	18
Tidal volume (ml/kg)	8.3	12.9	29,30
pO_2 (mmHg)	84–91	86–92	31,32
pCO_2 (mmHg)	35–44	37–46	30,31
HCO_3 (mmol/l)	17–29	20–25	30,31
Arterial blood pH	7.44	7.46	30,31
Arterial oxygen saturation (%)	85–87	97–99	33,34
Systolic arterial pressure (mmHg)	109–116	117–135	32,35
Diastolic arterial pressure (mmHg)	102	101–116	32,35

TABLE 5. HEMATOLOGICAL VALUES FOR SHEEP AND GOATS

Parameter	Typical Values		Reference(s)
	Sheep	Goat	
Packed cell volume (%)	27–45	22–38	20,36
Total red blood cells (10^6/ml)	9–15	10.5–13.5	20,36
Hemoglobin (g/dl)	9–15	8–12	36
Total white blood cells (10^3/ml)	4–12	4–13	36
Neutrophils (10^3/mm^3)	0.7–6	1.2–7.2	36
Lymphocytes (10^3/mm^3)	2–9	2–9	36
Eosinophils (10^3/mm^3)	0–1	0.05–0.65	36
Basophils (10^3/mm^3)	0–0.3	0–0.12	36
Monocytes (10^3/mm^3)	0–0.75	0–0.55	36
Platelets (10^5/mm^3)	2.5–7.5	3–6	36
Mean cell volume (fl)	28–40	16–25	36
Mean cell Hb concentration (g/dl)	31–34	30–36	36
Mean cell Hb (pg)	8–12	7–8	36
Blood volume (ml/kg)	57–66.4	70–70.6	37
Plasma volume (ml/kg)	46.7–61.9	51–55.9	22

2

husbandry

housing

The first point to be made about the housing of goats and sheep in a laboratory environment is that there is no one best way. Sheep and goats can adapt to a variety of housing situations, and when planning or designing housing for these animals it is important to keep in mind that they are social animals. A distinction must be made between housing small ruminants in a typical biomedical environment and housing them in an agricultural setting for food and fiber research. Guidelines exist for both environments and should be referred to for additional information.[38,39] The two sets of guidelines may differ on housing and husbandry issues, and if a discrepancy arises, the intended use of the animal determines which guidelines take precedence. Table 6 summarizes the recommendations for small ruminants used in biomedical research.

In addition to the aforementioned guidelines, the Animal Welfare Act has both general and specific requirements regarding the care and use of sheep and goats in research, education and testing.[40]

The enclosure immediately surrounding the animal is called the primary enclosure. In general, the primary enclosure must be soundly constructed and well maintained. It needs to be made

TABLE 6.   HOUSING REQUIREMENTS FOR ADULT SHEEP AND GOATS USED IN BIOMEDICAL RESEARCH AND TESTING.[38]

Number of Animals per Enclosure	Weight, kg	Floor Area per Animal, m²
1	<25	0.9
	25–50	1.35
	>50	0.8
2–5	<25	0.765
	25–50	1.125
	>50	1.53
>5	<25	0.675
	25–50	1.017
	>50	1.35

of materials that can be sanitized or replaced when worn or soiled, and must be free of sharp projections. The animal needs to be able to keep clean and dry in the primary enclosure, and needs ready access to food and water. There needs to be sufficient space in the primary enclosure to permit normal activity in an unrestricted manner. If primary enclosures are located outdoors, they must protect the animals from extremes in weather and provide sufficient shade for all the animals in the enclosure.

In the laboratory environment, goats and sheep are typically housed in runs or pens (Figure 3). The type and texture of the flooring material is important, because the animals must have adequate footing to prevent falling and subsequent injuries. If elevated, the flooring should be made of material that will allow the animals to ambulate normally while permitting feces and urine to pass through the perforations in the flooring material. Elevated rubber coated flooring works well, as it provides a way to keep the feces and urine away from the animals. The flooring, whether elevated or not, must provide sufficient traction for the animals, as sheep or goats may become excited or anxious when disturbed and may slip and fall if the flooring is slick. This may also be a concern for the personnel in the housing area, especially after the area has been cleaned with water. Pens for sheep and goats often have drains in them to permit the area to be washed with water and disinfectant.

Fig. 3. Typical pen housing for sheep or goats.

Bedding material such as straw, wood shavings, or sawdust may be used in the pens, and must be removed from the floors prior to washing, to prevent the drain lines from becoming plugged.

In certain circumstances, outdoor housing provides an attractive alternative for small ruminants on experimental protocols (Figure 4). Although outdoor housing offers a number of advantages over indoor housing, particularly with regards to per diem costs, issues such as fencing, predator control, and parasite control are critical in these units. It is also important to consider the practicalities of exposing valuable experimental animals to potential extremes of weather, as well as to possible scrutiny by local animal rights groups.

For indoor units, the walls of the primary enclosure should be non-porous and easy to sanitize. Tiles or unpainted walls are preferred because of the risk of accidental lead poisoning from ingestion of paint; this is more likely to occur with goats.[8]

There are numerous materials that are acceptable for pen construction, as long as sanitizability, durability, and safety are kept in mind. In many laboratory animal facilities, runs that

Fig. 4. Outdoor housing for small ruminants. It is important that animals have access to shade and shelter from wind and rain.

were originally constructed for housing dogs can be used for housing small ruminants with little or no modification, if the housing area is not being used for dogs. The runs need to be large enough to meet the space requirements of the animal, and must provide adequate footing. Goats are more inquisitive animals than sheep, and special attention needs to be given to potential danger areas, such as small gaps that could serve as a catch point for goat hooves. Also, if the animals are group housed in large rooms, any items that could injure the animals or be ingested need to be removed or sealed or made inaccessible. Examples of such items include water faucets that could be nuzzled open, clipboards or thermometers hanging on the walls, or a loose drain cover.

As mentioned earlier, sheep and goats are social animals for whom isolation is stressful. This may be more of a problem if the animals were group-housed at the supplier, and then placed in isolation upon arrival. As much as possible, sheep and goats should be able to see and hear other animals of their species. Ideally, sheep and goats will be housed separately, but a more important consideration is separation by source. This is especially important when a new source of animals is being utilized. If animals are group housed, close observation is needed initially to ensure that the animals are compatible and no fighting occurs. It is not recommended that sheep and goats be group-housed together, for social, behavioral, and managemental reasons.

If research protocols mandate that the animals be housed in metabolic units, it is important that there be sufficient space for normal postural adjustments (Figure 5). The animals should be acclimated to the metabolic units to minimize stress associated with housing in these units. Also, the animals should be kept in these units the minimum length of time necessary to successfully complete the experiment. Increased frequency of monitoring by research and care personnel should occur for animals that are in metabolic units, especially if monitoring equipment such as electric leads or catheters are attached.

Fig. 5. Metabolism cage. The animal should have sufficient room to make normal postural adjustments while confined in the metabolism cage.

environmental conditions

The foremost purpose of the laboratory animal facility is to maintain a controlled and stable environment to minimize non-experimental variables. Variations in the environment can cause physiological alterations in the animals, and potentially adversely affect the experiment. In contrast to most laboratory animals, which are obtained from commercial vendors, housing conditions for small ruminants are likely to be quite variable between suppliers, and also may differ significantly from a typical laboratory animal housing environment. This change may be an additional source of stress to the animals, and reinforces the importance of providing an adequate acclimation period for the animals upon arrival.

The following variables should be considered when setting up laboratory housing for small ruminants:

Temperature. Sheep and goats can tolerate a wide range of ambient temperatures; however, temperature adaptations take time and involve physiological changes that can influence research. Temperatures should be stable within a few degrees, and dry-bulb temperatures should fall in the range 61° to 81°F.[39]

Illumination. Adequate lighting is important to ensure normal circadian rhythms in sheep and goats, and is also needed to allow normal husbandry and adequate veterinary care. A light intensity of 300–325 lux about 3 feet from the floor is usually adequate to meet both requirements. Commonly, animals are provided 12 hours light and 12 hours dark through the use of automatically controlled light. Reasons to alter the light cycle or intensity include experimental justification or a desire to induce estrus in the animals for breeding.[41] Windows have typically been considered inappropriate in rooms housing laboratory animals; however, in recent years, they have been considered acceptable or even desirable for the purposes of environmental enrichment. Issues to consider when installing or utilizing windows in animal rooms include temperature, photoperiod fluctuations, and security.

Ventilation.   Adequate ventilation is important in the animal room to maintain stability of gaseous and particulate contaminants, and to dilute and remove heat, humidity, and odors generated by the animals. The ventilation must be well distributed throughout the room to effectively maintain the air quality and to minimize drafts. Ventilation rates of 10 to 15 air changes of 100% fresh air per hour are commonly employed for rooms housing sheep and goats. Recirculation of room air is not advisable due to the increased risk of disease, and potential concerns over the build-up of pheromones.

Humidity.   Relative humidity levels should be maintained in the 30 to 70% range for animals housed indoors. Humidity levels can vary, depending on the number of animals per pen or room, the choice of bedding material, the frequency with which the bedding is changed, and the method of cleaning (wet vs. dry).

Noise.   Sheep and goats are easily startled by loud noises. Housing areas for small ruminants should be located away from equipment and animals that generate significant amounts of noise (cage-washing area, heating/ventilation/air-conditioning equipment, dogs, swine, and nonhuman primates). Additionally, sheep and goats can vocalize enough to distract or interfere with other animals, especially when newly received. Sound dampening baffles are available and can be used to attenuate sound levels in sheep and goat housing areas.

environmental enrichment

Environmental enrichment is a phrase popularized in recent years, and refers to attention to the **psychological needs** of the animal by altering the environment. As sheep and goats are gregarious, group housing the animals is recommended as a form of environmental enrichment, although the gender and reproductive status of the animals should always be taken into account. If there are practical reasons why animals cannot be group housed, every attempt should be made to ensure visual and/or auditory contact with conspecifics. Contact with humans during routine husbandry procedures, or at other designated times, can also serve as environmental enrichment. The provision of natural light has also been suggested as a means of enriching the animals' environment.

While there are few definite recommendations in the literature for environmental enrichment of sheep and goats, common examples from other species include use of radio or television/video tapes, introduction of novel objects that can be moved or manipulated (e.g., soccer ball). Food treats may also be used, and efforts should be made to provide these in a manner that allows the animal to express its normal feeding behavior (grazing at floor level for sheep, browsing off the floor for goats). As mentioned previously, goats are especially inquisitive and will eat almost anything, so it is important to avoid objects that may lead to foreign body ingestion and impaction.

nutrition

Sheep and goats are **herbivores**, and when indoors or in areas where they do not have access to pastures, they should have access to an alternative source of roughage. This usually consists of baled hay, although alfalfa cubes or silage are also suitable. These materials can be fed in a trough or slot feeder, or a hay net can be used to keep the material off the floor. The availability of pasture is an option for small ruminants, though it is recognized that pasture is often not readily available in locations where these animals are used in biomedical research. If pasture access is available, the routine for monitoring and treatment of internal parasites must be more intense than for those animals in complete indoor housing (see Chapter 4). It is normal for sheep and goats to ruminate for 6 or more hours daily, and observing animals chewing their cud is an indication that they are eating and displaying normal digestive functions. In addition to providing normal feeding behavior, pastures also provide trace minerals, which need to be supplemented for non-pastured small ruminants.

Common nutritional problems in small ruminants include selenium deficiency, pregnancy toxemia, and urolithiasis; these will be discussed in greater detail in Chapter 4. Other potential nutritional problems include rumenal acidosis, bloat, thiamine deficiency, copper deficiency, and copper toxicosis. Rumenal acidosis is most usually associated with ingestion of excessive quantities of grain, while frothy bloat typically develops if animals are exposed to lush legumes or fresh grass.[42] Thiamine deficiency can cause polioencephalomalacia, which manifests as incoordination and possibly blindness.[43] Copper deficiency can cause "swayback" in lambs, but occurs much less frequently than copper toxicity. Sheep are at risk for copper toxicity because they lack the enzyme that degrades hepatic copper stores; copper toxicity develops when sheep are fed rations containing excessive amounts of copper.[44] Sheep must never be fed cattle rations, for this very reason.

Grain or pelleted feed is used to meet the energy needs of the animal, and these needs vary greatly depending upon the animal's age, intended use, and reproductive status. Animals housed indoors have low energy requirements if ambient temperatures are kept near the animal's thermoneutral zone, whereas outdoor housed animals may have greatly increased energy requirements. The nutritional requirements for sheep and goats are similar, and have been determined and published by the National Research Council.[16,45]

Water.   Water can be supplied to the animals by automatic systems (nipples or cups) or with buckets. The personnel taking care of the animals may need to help train the animals to the watering technique depending upon the type of system to which the animals have become accustomed. Ruminants, including sheep and goats, consume large amounts of water daily, so plentiful supplies of fresh potable water should be available. If the water is supplied in buckets or pans, these containers need to be elevated to keep the water clean and need to be secured to the wall so they are not kicked around in the pen. The water is consumed throughout the day, so it should be available *ad libitum*.

Feed Storage.   If commercially available feed is purchased, the bags should be stored in a dedicated feed storage area. Storing feed at cool temperatures will increase the longevity of nutrients (Figure 6). Hay storage is more problematic, due to the bulk and nature of packaging. In contrast to the stringent nutrient and packaging requirements of most feeds utilized in a laboratory animal environment, hay is typically less controlled during manufacture and storage. Hay samples can be analyzed for nutrient content, and it is recommended that the source and storage of hay from the supplier be checked to prevent introduction of pathogens into the laboratory animal facility.

Fig. 6. Storage area for feed. Storage at low ambient temperatures will optimize the stability of pelleted diets.

bedding

Bedding materials need to be provided for the animals, and the most common types are straw, sawdust, or wood shavings. If the removal of waste bedding is a problem, rubber mats can be used to provide a non-slick, cushioned surface in primary enclosures.

Clean bedding should be stored away from the main housing area or, if space is limited, in a subsection of the housing area. Bedding materials for sheep and goats are not typically stored in the same areas as bedding materials for other laboratory animals, due to the source and type of materials being used.

sanitation

Sanitation refers to the procedures and schedules for cleaning. All materials, surfaces and equipment must be considered when developing the sanitation program, as dirt and dust can provide a potential growth area for microorganisms. The frequency of cleaning, the agents or materials utilized, and the methods must all be taken into consideration.

Frequency

The primary enclosure should be cleaned daily, or if experimental conditions prevent daily cleaning (e.g., metabolism cage housing), the unit should be designed such that excreta can be easily removed from the area. The room and pens can be cleaned by multiple methods, including scooping up the used bedding, scraping or washing the area, or vacuuming the waste material with a high-capacity HEPA-filtered vacuum. Due to the dusty nature of the hay and bedding material, the secondary enclosure (the room containing the pens or runs) needs to be cleaned one to several times per week, depending on the housing density, type of bedding materials, and room ventilation rates.

Methods

Both the primary and secondary enclosures need to be regularly sanitized. Sanitation involves the removal of grossly visible debris such as bedding, feces, and spilled feed. It also involves the removal of surface films of oil and dirt that may not be visible during casual inspection. Once the debris has been removed, the area can be effectively disinfected. Disinfection refers to the reduction or elimination of harmful microorganisms. For disinfection to be effective, surfaces must be nonporous and free of gross contamination.

surface cleaning

Grossly visible debris is generally washed off with running water. Brushing and rinsing with water may be sufficient for cleaning; however, detergents are usually used to loosen or dissolve debris and to lift the oils and organic matter that often coat animal enclosures.

disinfection

A variety of disinfectant chemicals are available from a number of vendors (see Chapter 6). The products used should be formulated specifically for the laboratory animal environment. These chemicals often have detergent properties and their use accomplishes both cleaning and disinfection. Any application of chemicals to pens or equipment should be followed by thorough rinsing with water to minimize exposure of the animals to potentially harmful chemicals; the sheep and goats should be removed from their enclosures and protected from the application of disinfectants or hot water during the cleaning process.

For removable items (e.g., transport boxes, feed racks), disinfection may be accomplished with hot water of 180°F in a cage washer.

> **Note:** Individuals performing these procedures should wear protective equipment to minimize the risk of chemical or thermal injury to the skin, eyes, and nasal passages. Hearing protection is also advisable, due to the noise generated by the animals and the cleaning.

Quality Control

It is important that the sanitation effectiveness be monitored. On a daily basis, visual and olfactory inspections are important tests for monitoring overall efficacy. A more objective program, such as sanitation temperature or microbiologic monitoring can be employed.

sanitation temperature

Test tapes are available that will undergo a color change when exposed to temperatures consistent with disinfection. The test tape is applied to objects that are being washed in the cage washer. Monitoring with temperature tape is usually performed on a weekly basis, and records of acceptable cage wash temperatures should be maintained within the animal care facility.

microbiological monitoring

Since a major objective of sanitation is removal of microorganisms and the dirt that shields them, the best assay for sanitation efficacy involves bacterial culture of equipment and room surfaces. Surfaces to test include animal pens, cages, watering and feeding equipment, and any other equipment that undergoes sanitation. The surfaces can be sampled with a swab or Rodac plate and incubated. Moderate or heavy bacterial growth indicates weakness in the disinfection process. Gram-negative, rod-shaped organisms are the hallmark of fecal contamination and when detected indicate a need to re-evaluate procedures.

transportation

Transportation arrangements for sheep and goats must meet the requirements of the animals, and depend on time in shipment, numbers of animals, ambient temperatures and humidity. If the animals are likely to be exposed to extremes of weather in transit, climate-controlled transportation should be provided. The appropriate documentation must be filed and accompany the animal shipment. The vendor should be USDA-licensed if selling to a research facility, and the transporter should be registered with the USDA.[46,47]

> **Note:** The Animal Welfare Act should be consulted any time small ruminants are shipped.

Briefly, important aspects of transportation include records, the shipping container, provision of food and water during transit, and environmental factors.

Records. Official documentation for the transportation of small ruminants must comply with the requirements of the Animal Welfare Act.[40] Shipments of animals to be transported by commercial carriers must be accompanied by a health certificate signed by a licensed veterinarian. The health certificate must have been prepared within 10 days of transportation, and must indicate that the animals appeared free of any signs of disease or injury that would endanger either the animal or in-contact humans. An additional certificate should indicate who is sending the animals, and give the USDA-assigned identification numbers for each animal in the shipment. Official documentation should also include instructions for feeding and watering the animals during transit (see below).

Shipping Container. The shipping container must contain the animal in a safe and secure manner. It should be constructed of a durable material that is resistant to damage and that will not injure the animal. The container must be labeled to indicate that it contains live animals, and to identify which side should face up. Sufficient space should be allowed to allow each animal to make normal postural adjustments during transit, and to protect the animals from urine and feces. Containers should be designed to allow adequate ventilation.[47]

Food and Water. The need for food and water depends on the length of time and the conditions of shipping. Food and water must be provided within 4 hours of delivery to a commercial carrier,[47] and specific instructions for feeding and watering must accompany the shipment. As a minimum, sheep and goats must be fed at least once daily and provided with water for at least one hour on two occasions per day. Food and water receptacles should be secured inside the shipping container for this purpose.

Observation. The regulations of the Animal Welfare Act call for observations every four hours, except during air travel, when it is not possible to do so.[40] Observations should include assessment of the general condition of the animals, as well as assessment of the environment in and around the shipping container. Animals developing health problems should receive prompt veterinary care.

Environment. The environment surrounding the small ruminant is of great importance. The ventilation system should provide fresh air with minimal drafts, and extremes of temperature should be avoided. Any abnormalities noted during transit should be communicated to the attending veterinarian as soon as the shipment reaches its final destination.

record keeping

There are a variety of records that need to be kept, and numerous reasons for these records. Records help to standardize the care and treatment of the animals, and are required by regulatory agencies. Records should be maintained such that they are easily accessible and understood by the users, yet they need to be protected from damage by the animals and cleaning routines.

USDA Records. All small ruminants must be uniquely identified in a USDA approved manner. This identification must be traceable from the time of acquisition until at least 3 years after disposal, and should consist of either an ear tag, collar, tattoo, or implanted microchip (Figure 7).

Sheep and goats obtained from external sources must have acquisition records. These records must identify the source of the animal, the USDA registration or license number of the source, the date of acquisition, the animal's USDA number, and a complete description of the animal.

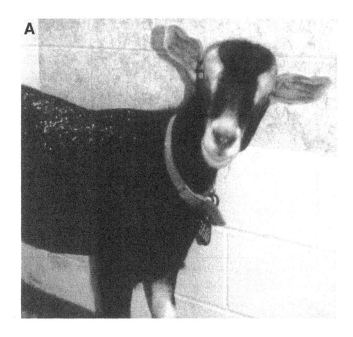

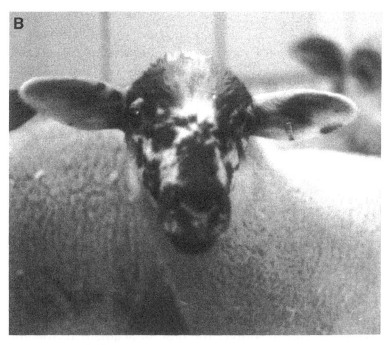

Fig. 7. Identification systems for small ruminants **(A)** collar and plastic tag; **(B)** metal ear tag.

Note: An annual report of research must be filed with the USDA before December 1 each year, detailing sheep and goat use during the previous fiscal year.

Health Records.   Adequate veterinary care requires an animal history when problems develop. Records need to be kept on each animal from the time it enters the facility until the time of euthanasia and disposal. Additionally, records of experimental manipulations need to be recorded. Records of operative procedures and pre- and postprocedural care are necessary so it is possible to formulate a diagnostic plan if a health problem develops.

Census Records.   Information concerning the number of animals in individual rooms is helpful in planning the daily workload. Census information is also used when recovering costs through a per diem system, and when reporting animal use to regulatory agencies

Work Records.   Daily records of routine husbandry tasks carried out in animal rooms should be maintained. Basic relevant information includes food and water provision and intake, changing of pens or cages, temperature and relative humidity, and initials of person logging the information.

---

breeding

It is rarely practical for institutions to breed and raise small ruminants for use solely in biomedical research. However, institutions performing agricultural research with small ruminants usually have a breeding flock.

Reproductive Cycle

Reproduction in sheep and goats has been reviewed extensively elsewhere.[8,41,48] Normal values for a range of reproductive parameters are shown in Table 7.

TABLE 7. NORMAL REPRODUCTIVE PARAMETERS FOR SHEEP AND GOATS.

	Typical Values		
Parameter	Sheep	Goat	Reference(s)
Estrous cycle length (days)	15–19	18–24	8,48
Estrus (hours)	30–48	30–40	48
Timing of ovulation	Spontaneous 30–32 hr after onset of estrus	Spontaneous 30–36 hr after onset of estrus	48
Gestation length (days)	147–156	143–154	10,12,48
Average litter size	1–3	1–3	10,12
Average birth weight (kg)	3.5	2–3	10.12

- **Puberty** typically occurs between 6 and 10 months for sheep and 5 and 7 months for goats.

- Most breeds of small ruminants are **seasonally polyestrous**, undergoing estrous cycles in the fall and early winter. A number of sheep breeds have been developed which are capable of two breeding seasons (Dorset) or year-round breeding (Finns).

- The influence of light patterns on breeding behavior has been studied in detail in sheep. It is recognized that a pattern of shortening day length,[49] typical of the transition from summer into fall, is a potent stimulus for the development and activation of estrous in the female.

- Techniques have also been developed for **artificially stimulating** the onset of estrous cycles outside the normal breeding season; these include the use of progestagen sponges, pregnant mare's serum gonadotrophin (PMSG), light stimulation and melatonin.[49,50] The introduction of a vasectomized (teaser) ram or buck into the flock just before the start of the natural breeding season may be used to stimulate the onset of estrous in ewes and does.[41] Once the females are cycling, a fertile ram or buck can be introduced.

Estrus Detection

- Although estrus detection is practical in does, it is difficult in ewes, so it is common practice to simply allow the ram to run with the flock in order to ensure maximum fertility rates.

- The ram may be equipped with a harness (raddle) on which a colored marking device is mounted; when the ram mates with a ewe in estrus, a colored mark is transferred onto the fleece of the ewe. In this way, it becomes relatively easy to identify when ewes were mated.

- Since the ewes will only return to estrus activity if they are non-pregnant, it is also possible to identify ewes which are failing to conceive after mating; these ewes should be examined by a veterinarian, and if they are determined to be healthy, returned to the flock and monitored carefully.

Pregnancy

- **Gestation** length is approximately 147 days in both sheep and goats, although there is variation between individual animals and, in particular, between different breeds.

- **Pregnancy detection**. In a commercial setting, many farmers rely on failure to return to estrus as an indicator of a successful mating. However, pregnancy detection can best be achieved by ultrasound examination of ewes and does at >45–50 days of gestation.[51,52] Biochemical assays for serum progesterone (>21–24 days after service) and milk estrone/estrone sulfate (>50 days after service) have also been used in goats.[53]

- Enlargement of the abdomen and/or mammary glands should not be taken as definitive evidence of pregnancy in goats because of the risk of **pseudopregnancy** or "cloudburst".[8]

Parturition

- In both sheep and goats, parturition occurs more commonly in the evening or overnight than during daylight hours.

- Typical signs of first-stage labor include general restlessness, separation from the flock, and a reduction in appetite.

- Second-stage labor is evidenced by increased straining, usually with the animal in sternal or lateral recumbency. The cervix should have dilated at this stage. **Ringwomb**, a condition in which the cervix fails to dilate, is an important cause of dystocia in both sheep and goats.[48]

- Most ewes and does will deliver without assistance. Anterior presentation is normal in both sheep and goats, and any deviation from normal should be managed promptly; the uterine canal is relatively wide in both sheep and goats, and extensive manipulations are generally possible, allowing successful correction of the majority of malpresentations.

- If straining persists for more than two hours without apparent progress, the animal should be examined by a veterinarian.

- Third-stage labor involves the delivery of the placenta and the remains of the umbilical cord.

- Retained placenta, which is a relatively common problem in cattle, is rarely seen in sheep and goats.

- Corticosteroids can be used to induce parturition in sheep and goats.[54]

management of neonatal lambs and kids

- Like other newborns, the neonatal small ruminant is extremely susceptible to hypothermia, hypoglycemia, and infectious disease.[55]

- **Colostrum intake** is critical in the neonate. If the lamb or kid will not feed, a stomach tube should be passed and colostrum (from either the natural mother or a surrogate) administered at a dose rate of 50 ml/kg.

- Once they have received an adequate supply of colostrum, lambs and kids can be hand-raised on commercial milk substitutes.

- If animals are to be castrated, the procedures should be performed early in life (within the first week is preferable). A number of techniques can be used, including rubber rings, bloodless emasculators, and open castration.[8,56]

- Techniques for disbudding in kids have been described elsewhere.[11,57]

- Artificially, lambs and kids may be weaned after approximately 8 weeks; in the natural setting, they are weaned at about 5 to 6 months of age.

notes

management

---

regulatory agencies and compliance

Specific regulatory agencies and requirements may vary with locale. In the United States, the following are the primary organizations with regulatory oversight or accreditation responsibilities for programs of research, teaching, or testing involving small ruminants:

The United States Department of Agriculture (USDA)

- Oversight responsibility is described in detail in the **Animal Welfare Act.**[40]

- Specific regulatory requirements are described in the **Regulations of the Animal Welfare Act**.[46,47]

- Registration with USDA and adherence to USDA regulations is required by all institutions, except elementary or secondary schools, using small ruminants in teaching, biomedical research, or testing in the United States.

- Unannounced, on-site inspections by the **Animal Care** section should be anticipated at least once a year; more frequent visits are not uncommon.

The National Institutes of Health, Public Health Service (PHS)

- Oversight responsibility is described in the **Health Research Extension Act of 1985**.[58]

- Policy is described in the Public Health Service Policy on the Humane Care and Use of Laboratory Animals.[59]

- Adherence to PHS Policy is mandatory for all institutions performing research which is funded by the PHS.

- Principles for implementation of PHS policy are those described in the *Guide for the Care and Use of Laboratory Animals*.[38]

The United States Food and Drug Administration (FDA) and the Environmental Protection Agency (EPA)

- Policies are described in the document **Good Laboratory Practices for Nonclinical Laboratory Studies** (CFR 21 (Food and Drugs), Part 58, Subparts A-K; CFR Title 40 (Protection of Environment), Part 160, Subparts A-J; CFR Title 40 (Protection of Environment), Part 792, Subparts A-L).

- In general, standard operating procedures must be outlined, followed and carefully documented with detailed records.

- Adherence is required when using sheep or goats in studies performed as part of an application for research or marketing permits for drugs or medical devices intended for human use.

Association for Assessment and Accreditation of Laboratory Animal Care International, Inc. (AAALAC)

- AAALAC International is a nonprofit organization designed to provide peer review-based accreditation of animal research facilities.

- Accreditation is based upon the adherence to the principles described in the *Guide for the Care and Use of Laboratory Animals*.[38]

- Accreditation is voluntary.

In addition to the above regulatory agencies, state and local regulations may exist.

institutional animal care and use committee (IACUC)

The Institutional Animal Care and Use Committee (IACUC) is the basic unit of the animal care and use program. The USDA, PHS, and AAALAC require an IACUC at any institution using sheep in biomedical research, teaching, or testing. Important points regarding the composition of the IACUC include:

Number of Members. USDA regulations prescribe a minimum of three members, but for institutions receiving funding from the PHS a minimum of five members is required.

Qualifications of Members. The IACUC should include the following:

- A chairperson.

- A Doctor of Veterinary Medicine (D.V.M.) who has specific training or experience in laboratory animal medicine or science, and who is responsible for activities involving animals at the research facility.

- An individual who is in no way affiliated with the institution (apart from his or her involvement with the IACUC). Frequently, this role is fulfilled by members of the clergy, lawyers, or representatives from the local humane society, animal shelter, or branch chapter of a medical charity.

In addition to these mandatory requirements, PHS policy requires the following members:

- A practicing scientist with experience in animal research.

- One member whose primary interests are in a non-scientific area. This individual may be an employee of the institution served by the IACUC, or could be the same person as the non-affiliated member (above).

It is acceptable for a single individual to perform more than one function within the IACUC, as long as the total number of members meets or exceeds the minimum specified in USDA or PHS policies.

Responsibilities of the IACUC. A detailed description of the responsibilities of the IACUC can be found in the relevant USDA and PHS regulations.[46,47,59] In general, however, the IACUC is charged with the following:

- Review of protocols for activities involving the use of animals in research, teaching, or testing. Protocols must be approved by the IACUC before animal use may begin. A majority vote is required.

- Inspect and provide written documentation that the animal research facilities and equipment meet an acceptable standard.

- Assure that personnel working with laboratory animals (either as researchers or as animal care personnel) have received adequate training and are qualified to conduct research in animals.

- Assure that investigators have considered all available alternatives to the use of animals in painful or stressful procedures. The investigator must also provide documentation that the research that is proposed does not unnecessarily duplicate previous work.

- Assure that sedatives, analgesics, and anesthetics are used whenever appropriate.

- Assure that researchers use appropriate techniques in surgical procedures performed on laboratory animals. For survival surgeries, the principles of aseptic surgery must be followed (see Chapter 4). Aseptic techniques are not required for non-survival surgeries, but researchers should still use clean instruments and wear appropriate protective clothing.

- Assure that animals are euthanized by an appropriate technique (see Chapter 4).

---

sources of small ruminants

There are generally two sources of sheep and goats: in-house bred and commercially purchased.

1. **In-house breeding** of sheep and goats allows a high level of quality control, but is expensive in terms of management, labor and cost. The production of sufficient numbers of animals of the correct age and sex requires a large and intensively managed breeding colony. The basics of breeding were discussed in Chapter 2, but it should be realized that, for most applications, it is rarely cost-effective for institutions to maintain their own breeding colonies. Specific exceptions to this general rule might be centers performing large numbers of fetal surgeries (where transportation of pregnant animals is undesirable), and agricultural research centers with an active animal science program, where animal husbandry techniques are being taught.

2. **Commercial vendors** are the main source of small ruminants for research. These vendors may produce the animals through their own in-house breeding programs, or acquire them from markets, other dealers, or even commercial farms. Facilities where sheep and goats are purpose-bred for research are designated Class A dealers; in all other instances, the suppliers are licensed by the USDA as Class B dealers. Class A dealers are preferred wherever possible, since the medical and husbandry history of the animals is known, and they are more likely to be of a reasonably high genetic and temperamental uniformity. Class B dealers may be able to provide animals more quickly (since they are simply purchased from other sources), and are often less expensive than Class A dealers, but there can be significant problems with ill

health in these animals. These problems are compounded when animals from multiple sources (and with different backgrounds of disease and immunity) are brought together at the dealer's facility.

quarantine and conditioning

All small ruminants, regardless of source, should be quarantined and conditioned (acclimated) by the research facility prior to use in any experimental procedures. The exact length and nature of the quarantine/conditioning period is a matter of preference, but since the majority of infectious diseases of small ruminants have relatively short incubation periods, a minimum of two weeks is recommended for most applications. Longer quarantine periods may be necessary for animals coming from random sources and destined for survival surgical procedures, and shorter periods may be acceptable if the animals are to be used in non-survival procedures and will not come into contact with other small ruminants in the same facility.

> **Note:** Respiratory and enteric diseases are important causes of morbidity in sheep and goats. The incidence of these diseases in a research colony can be minimized by careful control of the source, transportation and quarantine of newly acquired animals.

The quarantine and acclimation process should be designed and supervised by the attending veterinarian. Typical components of the process include:

- Examination of the shipping documents to ensure that they are accurate and complete.

- Examination by the technical staff to check that the requirements of the order (age, sex, breed, number of animals) have been satisfied.

- Physical examination by a staff veterinarian to detect medical problems and to identify any pregnant animals.

This examination should include particular emphasis on the examination of the animals' teeth and feet (see Chapter 4). Routine foot care should be practiced in any facility using small ruminants in chronic studies.[10]

- If appropriate, serology, bacterial cultures, blood smears, fecal flotation and fecal culture may be required to assess past exposure to infectious diseases and potential for disease carriage. Serological tests for caprine arthritis-encephalitis virus (Chapter 4) are indicated for any goats scheduled for long-term survival procedures.

- The animal should be given a permanent identification by means of an ear tag, tattoo, freeze-brand, or implantable microchip.

- Vaccination against common ovine and caprine diseases (see Chapter 4). Animals from Class A dealers may have been vaccinated previously.

- Prophylactic treatment with anthelmintics and topical insecticides to eliminate internal and external parasites (see Chapter 4).

- Depending on the time of year, shearing may be considered for sheep intended for chronic studies; this will help to reduce contamination of the fleece and improve access to any implanted devices.

- Physical isolation from small ruminants which are already out of quarantine to allow any incubating disease to manifest. Each facility must develop its own guidelines, based on the anticipated use of the animal, and the source from which it was purchased.

- During the acclimation period, animals should be allowed to become accustomed to their new environment. They should be disturbed as little as possible, while still being exposed to all of the day-to-day activities of the animal care staff (feeding, cleaning, etc.).

occupational health and zoonotic diseases

In contrast with purpose-bred laboratory rodents and rabbits, which pose relatively little risk of infectious zoonotic disease, small ruminants purchased from commercial sources may harbor a number of potentially dangerous zoonotic agents. Personnel working with laboratory small ruminants can minimize the risk of disease transmission by following published guidelines for occupational health and safety.[60] In general, personnel should wear clean protective clothing (laboratory coat or coveralls) when working with sheep or goats. Occupational health guidelines for personnel working with small ruminants should cover the following issues:

Puncture, Bite, and Scratch Wounds. Sheep and goats are unlikely to bite under normal circumstances. However, care should be taken during dental examinations, particularly of the cheek teeth. If the animal does bite down on the examiner's fingers, serious damage can result. Injured fingers should be washed thoroughly in cold water, treated with topical antiseptic (e.g., dilute povidone-iodine), and covered with a sterile dressing. In view of the risk of Clostridial contamination from small ruminants, personnel working with small ruminants should be current with respect to tetanus immunization.

Back Injuries. Low-back pain is a significant cause of morbidity amongst the general workforce.[61] Personnel dealing with small ruminants are at high risk for both acute and chronic back injuries, and precautions must be taken to ensure that animal caretakers are familiar with safe practices for lifting heavy objects.[62] If animals must be lifted, and mechanical aids (such as an electric hoist) are not available, the load should be shared between at least two people.

Risk to Pregnant Women. Small ruminants, particularly sheep, pose a special risk to pregnant women. Zoonotic diseases such as toxoplasmosis, listeriosis, and chlamydiosis are capable of causing abortion (see below), and for this reason pregnant women should not be exposed to small ruminants.

Allergies.   Although there have been few specific reports of personnel becoming sensitized to small ruminants,[63] repeated exposure to animals and animal waste does pose a potential risk to personnel working with sheep and goats.[64] A greater risk is, however, posed by any hay or plant source used in the laboratory facility, since allergies to grass and plant pollens are widely recognized as environmental hazards.[65] As with other allergies, it is advisable for sensitive personnel to wear a face mask or fitted respirator, gloves, and a clean, washable laboratory coat or coveralls, or disposable apparel. Ideally, personnel who become sensitized to sheep, goats, or plant products should be reassigned to tasks which minimize or eliminate exposure to allergens. The advice of an occupational health specialist should be sought if reassignment away from small ruminants is not possible.

Experimental Biohazards.   Some studies may involve purposeful infection of animals with either known or suspected human pathogens. In such cases, standard operating procedures for handling biohazardous materials and infected animals should be established as part of the scientific review process; guidelines for the use of biohazardous materials in experimental protocols have been described in detail elsewhere.[66] Small ruminants pose an unusual problem in terms of their size, and special precautions must be taken to ensure the safe disposal of carcasses (see Chapter 4).

Zoonotic Diseases of Small Ruminants

Although humans are susceptible to a number of microbial infections which affect domesticated and wild animals, these are typically diseases which affect a broad range of hosts (e.g., salmonellosis, brucellosis, ringworm, and rabies). There are, however, a number of diseases which are seen predominantly in small ruminants but which are potentially zoonotic to man. This discussion will focus on five such diseases: orf, listeriosis, toxoplasmosis, chlamydiosis, and Q fever.

orf (contagious ecthyema; "sore mouth")

- Common, highly contagious pox viral infection of small ruminants.

- Clinical signs, which are typically seen in young lambs and kids, include crusty lesions around the mouth and nose (Figure 8). These lesions often become secondarily infected, and the animals may become lethargic and anorexic.

- The diagnosis is made on the basis of clinical signs and appearance of lesions. If confirmation is required, virus can be isolated from the skin lesions.

- No specific therapy is available. The disease will resolve with time, but supportive care (topical antibiotic cream if lesions are infected, hand feeding if unwilling/unable to feed on their own) may help. Severe cases may need to be euthanized.

- Control/eradication from a closed unit involves vaccination with a live attenuated vaccine.

- Affected animals should be isolated and handled with care. The virus is easily transmitted to humans by direct contact with lesions (Figure 9).[67] Human cases should be referred for medical attention (although the lesions will heal spontaneously).

listeriosis

- The causative agent is *Listeria monocytogenes*.

- The organism is able to survive for extended periods in feces and silage. Animals are infected when they ingest contaminated feed or water. The organism travels up the trigeminal and hypoglossal nerves, and gain access to the central nervous system.[42]

- In both sheep and goats, encephalitis is the most common clinical presentation. Clinical signs include a head tilt, circling, and cranial nerve deficits (including facial nerve paralysis). As the disease progresses, the animal becomes recumbent and develops convulsions.

- Early treatment with penicillin and oxytetracycline may be effective. However, in view of the zoonotic potential of this organism, it is probably wise to euthanize affected animals.

- Human infection with *Listeria monocytogenes* is a concern, particularly for patients who are immunocompromized.[68] Unpasteurized goat's milk and milk products (especially soft cheeses) are commonly implicated as a source of *Listeria spp.* in these cases.

- In pregnant women, transplacental spread can result in septicemia and fetal death and spontaneous abortion (miscarriage).[69]

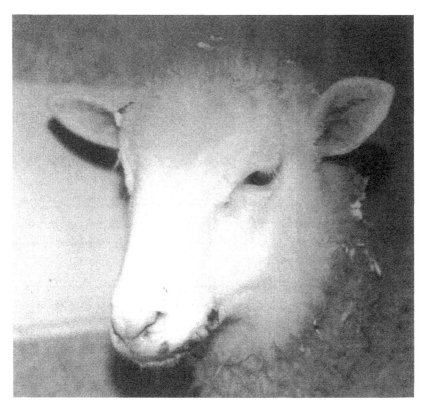

Fig. 8. Orf in a young lamb.

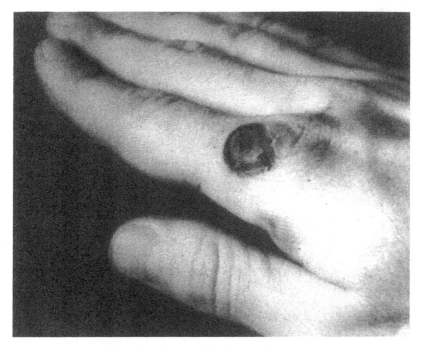

Fig. 9.   Case of orf in a human.

toxoplasmosis

- Causative agent is a protozoan parasite, *Toxoplasma gondii.*

- Cats are the definitive host for this parasite. Sheep and goats become infected through ingestion of parasite-laden cat feces. This often occurs from hay or straw that has been contaminated by cat feces.[70]

- In non-pregnant sheep, brief pyrexia may be the only sign of infection. If the animal is pregnant, infection spreads to the placenta and fetus.

- Infected ewes may abort or produce stillborn lambs. In some instances, lambs survive to term, but they are usually weak and die early in the postnatal period.

- Diagnosis is based on typical findings of white cheese-like flecks in the fetal cotyledons, and identification of the organism in fetal fluids and tissues.

- Asymptomatic human infection with toxoplasmosis appears to be common, but usually results in protective immunity.[71] However, infections may be fatal to immunocompromized patients,[68] and exposure during the second trimester of pregnancy can lead to abortion or birth defects, including chorioretinitis and mental retardation.[71]

chlamydiosis

- Causative agent is *Chlamydia psittaci*, the agent responsible for psittacosis in birds and enzootic abortion of ewes (EAE) in sheep. EAE is the number one cause of ovine infectious abortion in most regions of the United States.[42]

- *C. psittaci* is excreted in high numbers in the aborted fetus, fluids, and membranes. After ingestion by the ewe, the organism enters the animal's bloodstream and passes to the uterus.

- If the ewe is non-pregnant, there are no clinical signs and the ewe develops immunity.[72] If the ewe is pregnant, or becomes pregnant soon after infection, the developing fetus is aborted 60 to 90 days later.

- Aborting ewes are rarely sick.

- *C. psittaci* infection in pregnant women can lead to abortion.[73]

Q fever

- Rickettsial disease of ruminants caused by *Coxiella burnetii*.

- Occasional cases of abortion in sheep and goats have been linked to *C. burnetii*.

- The main significance of Q fever is as a zoonosis. In humans, the disease typically presents as flu-like symptoms of pyrexia, headache, myalgia, pneumonia, and hepatitis.[74]

- Strategies for controlling Q fever transmission within a laboratory environment have been published.[75]

- Suppliers of small ruminants should be encouraged to test their animals for seropositivity to *C. burnetii*,[76] and institutions should accept only seronegative animals for chronic studies.

veterinary care

basic veterinary supplies

The following basic veterinary supplies are useful for the clinical care of small ruminants:

1. Stethoscope.

2. Disposable syringes, 3-ml to 60-ml.

3. Disposable hypodermic needles, 20G and 18G, 1".

4. Dosing gun for anthelmintic treatment.

5. Electric clippers with both coarse blades (for rough clipping of thick wool) and fine blades (for close clipping).

6. Blood collection tubes with no additive (for serum) or with EDTA (for whole blood).

7. Clinical thermometer.

8. Water-based lubricant gel.

9. Arm-length, disposable plastic gloves.

10. Bacterial culture swabs.

11. Oxytetracycline aerosol ("purple spray").

12. Injectable antibiotics (long- and short-acting oxytetracycline, penicillin, and sulfonamides).

13. Sterile 50% dextrose solution.

14. Sterile calcium-magnesium-phosphorus-dextrose solution (CaMPD).

15. Hoof shears (sharp!).

16. Rope halter.

17. Elastrator and rubber rings for castrating and docking lambs.

Additional supplies, to supplement those listed above, should be purchased as required. For example, a dehorning iron will be useful for removing horns from young kids, and antibiotic pessaries are extremely useful around lambing time.

physical examination of small ruminants

The procedures for physical examination are similar for sheep and goats. The nature and scope of the physical examination will, however, be determined by the age (and sex) of the animal and the presence or absence of any obvious clinical signs. The following is an outline of the key points to be considered during physical examinations of immature and mature sheep and goats. These points are summarized in Tables 8 and 9.

Condition Scores
The use of condition scores has revolutionized the management of sheep as both breeding and meat animals.[77] A similar scheme may be used in goats.[57] The scoring system provides an extremely simple means of assessing an animal's nutritional status. The musculature on the dorsal aspect of the lower lumbar spine is used. The test is subjective, and if more than one observer will be scoring the animals during the course of a study, the observers should develop a standardized scoring scheme to ensure minimal inter-observer variations.

Head and Neck
In mature animals, particularly those arriving from Class B dealers, close attention should be paid to the condition of the

teeth. In mild cases, dental abnormalities, either from abnormal wear patterns or from dental disease, may cause few clinical signs and will only be detected by careful examination of the mouth.[9] In more severe cases, the animal may drool saliva from its mouth, be reluctant to eat solid foods, and perhaps resent examination of its mouth. In these cases, light sedation and the use of a gag may be necessary for direct visual observation of the problem teeth.

The **eyes** should be examined for evidence of irritation, discharge and corneal injury (either acute or chronic). The etiology and management of these problems will be discussed in detail below.

The local **lymph nodes** of the head and neck should be palpated in order to assess size, shape and consistency. Any animal showing signs of abscess formation in or around a lymph node should be isolated immediately (see later).

The **ears** should be assessed for evidence of irritation, hair loss (alopecia), and abnormal position. It is common practice in commercial farming operations to use ear tags for identification, and this can sometimes cause the animal to carry this ear slightly lower than the contralateral (untagged) ear; the abnormal ear carriage in this instance should not be confused with that seen in animals with a head tilt (see later). Parasites can also invade the outer ear and occasionally these will migrate down into the auditory canal, producing signs of head shaking and rubbing.

TABLE 8. CLINICAL EXAMINATION OF LAMBS AND KIDS

Examine:	For signs of:	Possible causes:
Umbilicus	swelling/redness	umbilical hernia, "navel ill"
Joints	swelling/pain/deformity	joint ill, contracted tendons
Eyes	abnormal lid margins	entropion/ectropion
	redness	conjunctivitis, corneal ulceration
Anus	patency	imperforate anus
	discharge	diarrhea
Mouth	soft palate	cleft palate
Lungs	abnormal sounds	pneumonia
Feces	diarrhea, blood	parasites, bacterial infection

TABLE 9. CLINICAL EXAMINATION OF THE MATURE SHEEP AND GOAT

Examine:	For signs of:	Possible causes:
Lumbar spine	condition score	malnutrition, anorexia, parasitism
Teeth	malocclusion, caries	dental disease
Eyes	abnormal lid margins	entropion
	redness	conjunctivitis, corneal ulceration
Feet	hoof overgrowth	inadequate foot care
	foul odor, discharge	foot rot
Skin and coat	alopecia	wool slip, parasites
	pruritus	parasites, scrapie
Lungs	increased sounds	bronchitis, pneumonia
	decreased sounds	pleural effusion, pneumonia
Abdomen	increased sounds	enteritis, overeating
	decreased sounds	GI stasis and/or obstruction
	distension	pregnancy, bloat
Superficial lymph nodes	increased size	inflammation, infection
Vulva	discharge	vulvitis, vaginitis, metritis
Mammary glands	heat, pain, swelling	mastitis
	fibrosis, thickening	chronic mastitis
Testicles	swelling	orchitis
Penis, prepuce	discharge, pain	balanitis, preputial infection
Feces	diarrhea, blood	parasites, bacterial infection

Thorax and Abdomen

Apart from a general examination of the hair coat, close attention should be paid to examination of the **respiratory system**. Signs of respiratory disease may include an increased respiratory rate (tachypnea), an increased respiratory effort (dyspnea), or both. Similar signs may be seen with a variety of systemic diseases in which the respiratory system is not the primary site of involvement, so it is important to auscultate and percuss the lung fields with a stethoscope in order to identify abnormal sounds and/or areas of tissue consolidation, both of which may be indicative of pulmonary disease.

Examination of the **cardiovascular system** should include routine auscultation of the thorax (to identify heart sounds), palpation of the peripheral pulse (usually at the femoral artery in femoral triangle/groin area) and an assessment of the animal's peripheral circulation (by examining color and capillary refill at mucous membranes).

Palpation of the **gastrointestinal tract** is extremely difficult in anything but newborn ruminants. Nevertheless, the abdomen should be examined visually (to look for evidence of asymmetric swelling, perhaps suggestive of gaseous distension) and auscultated (to identify normal sounds of stomach and intestinal motility). Fecal samples, either gleaned from the floor of the stall or removed by gloved hand from the animal's rectum, should be examined for color, consistency, and odor. Abnormal fecal samples should be submitted for routine fecal parasite screening and possibly bacteriology.

Examinations of the **urogenital tract** in the male should involve gentle palpation of external structures (penis, testes and epididymes) for evidence of abnormal size, shape or consistency. Acute inflammation of these structures typically presents as swelling, with associated heat and pain, whereas chronic disease more typically produces firm, non-painful swelling or atrophy.

In the female, the mammary gland, vulva and distal vagina are amenable to visual examination and gentle palpation. Acute infections of the vagina and/or vulva usually present with signs of vulval redness, swelling, and discharge (usually purulent). In the early post-parturient period, it is often difficult to distinguish vaginal/vulval infections from infections of the uterus (metritis), since in each case there will be a discharge. However, as a general guide, animals with metritis usually have (a) heavier vaginal discharge and (b) signs of systemic illness, including pyrexia and anorexia.

Acute inflammation of the mammary gland (mastitis) typically produces signs of local inflammation (with redness, heat and swelling) and discomfort. In severe infections, evidence of systemic illness (pyrexia, anorexia, and perhaps shock) may be seen. The management of these animals will be discussed later. In the chronic stage, fibrosis of the mammary gland is typical; animals with chronic mastitis are usually unable to secrete milk from the damaged gland and should probably be eliminated from the breeding flock. However, these animals may be entirely suitable for other research or teaching applications.

Fore and Hind Limbs

Animals should be encouraged to move around the stall so that the observer can identify lameness and isolate it to a particular limb.

Detailed inspections of the limb should concentrate first on the **foot**, since this is a common site of injury/disease in small ruminants. Trauma and infection are the two most common causes of foot problems in both sheep and goats. Hoof over-growth is rarely a problem in animals less than one year of age, but in older animals it can lead to the development of pockets of debris under and around the sole of the foot. Routine foot care, including visual examination of the interdigital skin and trimming of the hoof wall, should be performed on a monthly basis in laboratory facilities. Recommended techniques for hoof trimming in sheep and goats are described elsewhere.[10,78]

Routine examinations of the fore and hind limbs typically involve a combination of visual inspection (for signs of trauma and deformity) and palpation (to assess ranges of joint motion, evidence of joint swelling and pain, etc.).

common clinical problems and their treatment

A complete description of the diseases which affect small ruminants is clearly beyond the scope of this book, and the interested reader is encouraged to consult standard veterinary texts on sheep and goat medicine.[8,42,79] However, in contrast with many other species of laboratory animals, sheep and goats are commonly purchased from random sources, rather than from barrier units with established disease surveillance measures. As a result, small ruminants arriving at the laboratory can bring in a variety of both clinical and subclinical diseases, placing both the other animals and the personnel in the unit at risk. The most efficient methods for controlling disease entry into the laboratory are to use a reputable supplier, to inspect every animal before it enters the unit, and to be prepared to reject any animal that shows signs of disease. Even with these precautions, however, some animals that appear healthy may subsequently present with clinical disease. The etiology, clinical presentation, diagnosis, and management of the most common diseases affecting laboratory small ruminants are presented below.

Pneumonia

Respiratory diseases are probably the single most important cause of morbidity and mortality in sheep and goat units. Although upper respiratory tract infections do occur, they are much less important than lower respiratory tract infections. A number of bacterial and viral pathogens are known to cause respiratory diseases in small ruminants, including:

pasteurellosis

- Causative agent is *Pasteurella haemolytica*, a normal commensal in the nasopharynx and tonsils of healthy sheep. Disease results if the animal's resistance is lowered by intercurrent infection or, most commonly, poor management.[42] Dusty conditions, high ammonia concentrations, nutritional deficiency, overcrowding, and transportation stress have all been implicated as factors which reduce an animal's resistance to disease.

- Three syndromes of Pasteurellosis are recognized: (1) septicemia and death in young lambs up to 3 months of age, (2) pneumonia in older lambs and adults, typically in the spring and early summer and (3) septicemia and death in fattening and market lambs, typically during the winter.[21]

- The first sign of an outbreak may be sudden deaths in the flock. Animals in close proximity may show signs of respiratory involvement, with coughing, dyspnea, anorexia, fever, and ocular/nasal discharges.

- Diagnosis should be based on (a) history of recent stress (transportation, surgery, shearing) and (b) necropsy examination of fresh carcass, with isolation of large numbers of *P. haemolytica* organisms from lungs and other tissues (Figure 10).

- If identified early, animals with respiratory disease should be isolated and treated aggressively with antibiotics (in the absence of results from bacterial culture and sensitivity testing, treatment with two doses of long-acting oxytetracycline at an interval of 4 days is often extremely effective).

- Although eradication of the organism is not feasible (since it is a normal commensal), prevention involves good colostrum management in neonates and minimization of predisposing factors wherever possible; improve ventilation, minimize stress of handling. etc.

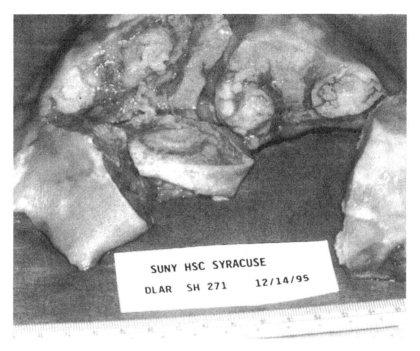

Fig. 10. Multiple lung abscesses caused by infection with *Pasteurella haemolytica.*

parasitic pneumonia (husk)

- Causative agent is a nematode, *Dictyocaulus filaria.*

- Clinical disease is much less common in sheep than in cattle, and is extremely uncommon in goats.

- Infections are acquired from contaminated pastures, so clinical disease is really only a problem in animals housed outdoors. However, animals arriving from random sources may harbor the parasite and can subsequently develop clinical disease.

- Larval burdens are highest in fall and winter. Larvae are resistant to cold weather and can infect young animals in the following spring.

- Clinical signs include coughing and dyspnea. Animals are rarely febrile.

- Diagnosis is based on clinical signs, along with identification of larval forms of *D. filaria* in a fresh fecal sample.

- Treatment with benzimidazoles or ivermectin is effective in clinically affected animals. Control depends upon prophylactic anthelmintic therapy (see Table 13) and avoidance of contaminated pastures.[21,80]

maedi-visna

- Causative agent is a slow virus (lentivirus) of the same family as the human immunodeficiency virus (HIV) and caprine arthritis-encephalitis virus (CAEV, see later).

- Clinical signs, which develop slowly and insidiously, can include respiratory disease, arthritis, neurological disease, and mastitis.[42]

- Most common clinical syndrome is ovine progressive pneumonia (OPP), which presents as gradual onset of exercise intolerance, coughing and dyspnea in older sheep.[81] Animals are rarely febrile.

- Diagnosis is based on clinical signs and pathological findings at necropsy (interstitial pneumonia and/or arthritis). Diagnosis is confirmed by agar gel diffusion test and virus isolation.

- The clinical signs of maedi resemble those of CAEV, and the virus can be transmitted to goats experimentally;[82] natural infections, however, do not appear to be very common in goats.

- There is no evidence that maedi-visna can be transmitted to humans.

Diarrhea

Diarrhea is predominantly a problem of neonatal and young (<1 year old) animals. Most cases of diarrhea are acute in onset, and many resolve spontaneously. However, infectious causes of diarrhea include:

neonatal diarrhea ("neonatal scours")

- Viral (rotavirus), bacterial *(E. coli, Clostridium perfringens type C)*, and parasitic *(Cryptosporidium spp.)* infections may be implicated.[21]

- *E. coli* scours are typically seen in lambs less than 1 week of age. Lamb dysentery (due to *Clostridium perfringens type C*) and *Cyryptosporidium* typically affect lambs 1 to 2 weeks old. Rotavirus infections are more common in lambs and kids 2 to 4 weeks old.

- Clinical signs include watery-to-pasty, pale-colored diarrhea (bloody diarrhea with lamb dysentery) and dehydration. If untreated, lambs will die quickly.

- Since it is extremely difficult to differentiate between causative agents on clinical signs alone, ancillary diagnostic tests are required.

- Samples of fresh feces should be submitted for bacterial culture (for *E. coli* and *Clostridium perfringens*) and virus detection (rotavirus). The oocysts of *Cryptosporidium spp.* can be identified in stained fecal smears.

- Fresh carcasses should be submitted for necropsy examination. Lamb dysentery produces severe ileitis, with necrosis and hemorrhage. Diagnosis is confirmed by identification of organism and/or toxin in feces.

- Symptomatic therapy involves fluid therapy (oral fluids for mild cases, subcutaneous or intraperitoneal fluids for more severe cases) and systemic antibiotics (Table 11). Even with early treatment, most animals with lamb dysentery will die.

- Prevention of neonatal scours centers on (1) improving hygiene, (2) ensuring adequate colostrum intake, and (3) vaccination of pregnant ewes and does against *E. coli* and *C. perfringens* (Table 12).

parasitic gastroenteritis (PGE)

- PGE is the most common cause of scouring and ill-thrift in lambs and kids from 2 weeks to 12 months of age. Clinical disease is rare in flocks maintained in an indoor setting.

- The most important gastrointestinal worms in sheep and goats are *Haemonchus spp.*, *Ostertagia spp.*, *Trichostrongylus spp.* and *Nematodirus spp.* Most infections are mixed.

- Immunity to parasites develops rapidly and older animals are therefore rarely affected by clinical disease.

- Adult ewes and does produce few infective larvae for most of the year, but fecal egg output increases dramatically in the periparturient period.[83] These eggs mature quickly in the spring and the infective larvae pose a significant threat to the new crop of lambs and kids.

- Infective larvae are able to survive winter on contaminated pastures. They therefore act as a significant threat to the following year's crop of lambs and kids.

- Clinical signs include ill-thrift, severe scouring, dull coats, dehydration, and death in young animals in the late spring and early summer. Subclinical infections may not be detected and are an important cause of reduced weight gain in growing lambs.

- Diagnosis is based on history and clinical signs, and confirmed by identification of significant numbers of eggs in fecal samples from affected and in-contact animals.

- For most flocks, control is achieved by a combination of (a) good pasture management and (b) routine anthelmintic treatments (Table 13).[84]

- Pastures grazed in the previous year are considered contaminated and, in an ideal world, should not be used for ewes/does and lambs/kids in the following spring. In practice, if these pastures must be used, regular anthelmintic treatment is required.

- Ewes and does are treated in the periparturient period (to prevent the periparturient rise in fecal egg counts) and for the first 6 weeks after they are turned out to pasture.

- As soon as they start to ingest significant quantities of grass (usually at about 1 month of age), lambs and kids should be dosed every three weeks, with treatments continuing until the end of spring, when the overwintered larvae die off.

coccidiosis

- A significant problem in most intensive sheep and goat operations, coccidiosis is caused by protozoan intestinal parasites of the genus *Eimeria*. At least 9 species of *Eimeria* are found in sheep, and 16 species have been reported in goats.[8] These parasites tend to be species-specific, and *Eimeria spp.* from sheep will not usually infect goats (or vice-versa).

- Infection typically passes from adult females to their offspring. Disease develops at times of maximal stress, including at lambing/kidding, weaning, after transportation, and in the presence of concurrent disease.

- Clinical signs early on include vague signs of ill thrift. Dark mucoid scours and soiling around the tail may be seen in some. Severe cases will progress to overt scouring, dehydration, emaciation, and death. Animals that recover may continue to be "poor doers" because of intestinal pathology.

- Infected lambs and kids then pass on the infection to other youngsters. Oocysts are excreted in feces and are resistant and remain viable on contaminated bedding (indoor flocks) or pastures (outdoor flocks) for periods of up to 12 months. Outbreaks typically start small and develop quickly.

- Diagnosis is based upon clinical signs. Examination of fecal samples is often unrewarding, since many animals excrete large numbers of non-pathogenic *Eimeria spp.* without signs of disease, and some animals may die without producing oocysts.

- Animals with clinical signs should be treated with systemic coccidiostats; oral or injectable sulfadimethoxine is effective.

- In facilities with a history of coccidiosis, prophylactic use of coccidiostats is recommended. Two doses of sulfadimethoxine (at 3 and 6 weeks of age) or continuous in-feed medication with lasalocid may be used (0.5–1 mg/kg PO in feed for up to 6 weeks).[85]

Clostridial Diseases

- Clostridial organisms are ubiquitous in the environment and cannot be eliminated.

- At least 10 forms of Clostridial disease are recognized in ruminants,[21] although their incidence and importance vary from region to region.

- Clinically, the most common diseases in sheep are lamb dysentery (see section on neonatal scours), pulpy kidney (caused by *Clostridium perfringens type D*) and tetanus (caused by *Clostridium tetani*). Clostridial enterotoxemia (*Clostridium perfringens type D*) is an important cause of mortality in kids.[8]

- Sudden death is the most common clinical sign. As with Pasteurellosis, outbreaks of clinical disease are usually associated with predisposing factors, such as trauma (tetanus) or overeating (e.g., pulpy kidney).

- Diagnosis is often made difficult because (a) the animal is found dead and (b) *Clostridium spp.* typically cause rapid putrefaction of the carcass.

- Prevention is based around the routine use of vaccines (Table 12). A popular choice in both sheep and goat units is a combination vaccine consisting of inactivated toxoids of *Clostridium perfringens types C and D* and *Clostridium tetani*.

- All breeding animals should be vaccinated. For pregnant animals, booster doses should be given within 4 weeks of lambing/kidding in order to ensure adequate transfer of colostral antibody to the neonates.

- Lambs and kids should be vaccinated at about 8 to 12 weeks of age, when maternal antibody levels fall. If they are to be kept for more than 4 months, they should be vaccinated again 4 weeks after the first dose.

Lameness

"scald" and foot rot

- Causative agents are *Fusobacterium necrophorum* and *Bacteroides nodosus*.

- *F. necrophorum* has many hosts and is considered a ubiquitous environmental contaminant. *B. nodosus* is carried on feet, and is capable of living within the environment for only limited periods of time (up to 2 weeks).

- *F. necrophorum* infection of the interdigital skin is known as "scald"; scald is easily cured by topical treatment with zinc sulfate.[86]

- Concurrent infection with *F. necrophorum* and *B. nodosus* produces "foot rot", which is much more severe and much more difficult to treat. Infection spreads from the interdigital skin and tracks up between the sensitive and insensitive laminae of the hoof wall (Figure 11).

- Grey-black, foul-smelling caseous material builds up between the laminae. If left unchecked, virulent strains of *B. nodosus* will cause complete loss of portions of the hoof wall.

- Clinical signs include lameness which, if both front feet are involved, can result in animals kneeling down on their carpi.

- Treatment involves trimming back affected hoof wall, removal of caseous debris, dipping of affected feet in zinc sulfate, and topical spraying with a tetracycline aerosol.[78,86]

- Severe cases of foot rot may also require systemic antibiotic therapy (penicillin-streptomycin, 20 mg/kg IM).[21]

- Control depends upon elimination of *B. nodosus* from the flock. Although vaccines have been developed, it is probably more prudent to cull problem animals.

- In a closed flock without evidence of foot rot, care must be taken to prevent re-introduction. All new arrivals should be screened and, ideally, should be purchased from clean sources.

Fig. 11.   Typical signs of foot rot in sheep.

septic arthritis ("joint ill")

- Most common in young lambs, as a result of systemic bacteremia or septicemia.

- Causative agents include *Streptococcus spp.* and *Escheriria coli.* Infection enters through the umbilicus (see navel ill) or by ingestion.

- Lameness in older sheep and goats may be the result of infection with *Erysipelothrix rhusiopathiae* and *Mycoplasma spp.*, respectively.

- Multiple joints are usually affected (polyarthritis); common sites include the stifle (Figure 12), hock, or carpus.

Animals are hunched and reluctant to move around. Young lambs may show evidence of systemic illness.

- Aggressive therapy with systemic antibiotics is indicated in all cases, but the prognosis is guarded. Potentiated amoxicillin is useful in lambs, and penicillin G is a good choice for older animals with *E. rhusiopathiae*,[21] while tetracyclines and tylosin have been advocated for Mycoplasmal arthritis in goats.[8]

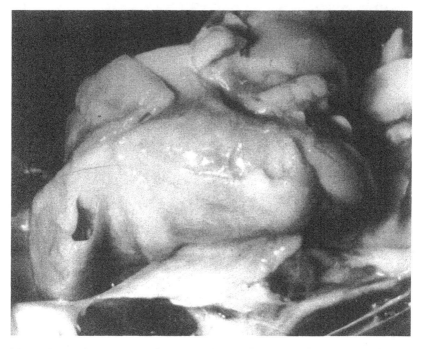

Fig. 12.   Fibrinous arthritis in the stifle joint of a sheep infected with *Chlamydia psittaci.*

caprine arthritis-encephalitis virus

See below.

Disorders of the Eye[87]

conjunctivitis

- Conjunctivitis is an extremely common disease in both sheep and goats.

- Unilateral disease is typically caused by trauma or non-specific infection. Bilateral disease is more often linked to specific infectious etiology, such as *Mycoplasma conjunctivae, Branhamella ovis, Streptococcus spp.* or *Chlamydia psittaci.*[88] Flies may transmit the infection between animals.

- Most commonly it is caused by superficial bacterial infection, leading to irritation, conjunctival congestion, blepharospasm, and ocular discharge.

- Severe cases may develop corneal ulceration, pannus formation, and corneal edema. If left untreated, these animals may lose sight in the affected eye.

- Diagnosis is usually based upon clinical signs alone. Fluorescein should be applied to affected eyes to determine whether the corneal surface is damaged. In young lambs, the eyelids should be checked for evidence of entropion (see below).

- Uncomplicated cases of conjunctivitis respond well to topical applications of tetracycline or cloxacillin ointment. Frequent dosing (up to four times daily) is required for maximum effect.

- If corneal ulceration is confirmed, aggressive treatment with topical and perhaps a subconjunctival injection of antibiotic (long-acting oxytetracycline) is indicated. Preparations containing steroids should be avoided until the ulceration has healed.

entropion

- Common condition of newborn lambs and kids.

- Clinical signs range from minor ocular irritation to severe blepharospasm, conjunctivitis and corneal ulceration. There may be uni- or bilateral involvement of the eyes.

- Mild cases of entropion, in which inversion of the margin of the eyelid is the only problem, are usually managed by manual eversion of the eyelid.

- Severe cases of entropion, in which manual eversion is unable to restore the normal eyelid margin, should be corrected with surgical staples, sutures, stainless-steel wound clips, or injection of long-acting antibiotic or liquid paraffin into the skin just below the margin of the lower eyelid.[89]

- Conjunctivitis and corneal ulceration should be treated as described previously.

- Entropion is believed to have a genetic basis, associated with a recessive gene.[90] Rams siring lambs with a high incidence of entropion should be culled. Affected lambs should be identified and should not be retained as breeding stock.

Diseases of the Urinary/Reproductive Tracts

abortion

- The most common causes of infectious abortion in sheep are *Chlamydia psittaci, Campylobacter spp.*, and *Toxoplasma gondii*. These agents, all of which are zoonoses, were discussed in Chapter 3.

- Salmonellosis, which is also a zoonosis, is a sporadic cause of abortion in both sheep and goats.[42] Ewes and does with Salmonellosis usually have clinical signs of bacteremia/septicemia (pyrexia, possibly with diarrhea and evidence of endotoxic shock) at the time when they abort.[21]

- Aborting ewes and does should be isolated; any bedding contaminated with fetal fluids or membranes should be incinerated, and the stall should be cleaned and disinfected before being used for another animal.

- Vaccines against certain infectious causes of abortion are available (Table 12).

metritis

- Metritis, inflammation of the uterine lining, is most often associated with bacterial contamination at or around the time of parturition.

- Mild cases of metritis present with a white or yellow purulent vaginal discharge. In severe cases, the ewe or doe is systemically ill, with anorexia and pyrexia.

- Treatment involves the use of antibiotics; mild cases may be treated with intrauterine pessaries or antibiotic-containing uterine flushes, but severe cases will require both local and systemic treatment.[8]

- Tetanus prophylaxis should be confirmed in all cases.

mastitis

- Mastitis, inflammation of the mammary gland or udder, can be a problem in both ewes and does.[91]

- Mastitis can be classified as either subclinical (no visual evidence of infection) or clinical (animal shows signs of disease). Clinical disease may be acute or chronic.

- Clinical mastitis is most often seen in breeding animals.

- Mild acute mastitis produces flakes or clots in the milk, often with only mild swelling of the mammary gland.

- Severe cases of acute mastitis involve sick animals, with hot, swollen, red udders, and abnormal milk (see Figure 13). The ewe or doe may be pyrexic (temperature 40°C) and anorexic.

- Chronic mastitis involves a persistent udder infection that is most often subclinical, but which has the capacity to flare up and cause episodes of acute disease. Animals with chronic mastitis usually have palpable lumps or fibrous thickening in the mammary tissue.

- Clinical cases of acute mastitis require prompt treatment with antibiotics. Mild cases respond well to local (intramammary) antibiotics,[21] while more severe cases should receive both local and systemic antibiotics. Fluid therapy and anti-endotoxic therapy (e.g., flunixin meglumine, 1.1 mg/kg IV or IM) may be indicated for animals presenting with signs of endotoxic shock.

- Animals with chronic mastitis should be eliminated from the breeding colony.

- In a facility where sheep and goats are purchased from class B dealers, care must be taken to identify animals with chronic mastitis. These animals may be suitable for many studies, particularly those involving non-survival surgeries, but the risk of an acute episode of mastitis may be unacceptable for researchers planning long-term survival procedures.

- Does with clinical evidence of mastitis should be checked for CAEV (see below). Ewes with clinical evidence of mastitis should be checked for ovine progressive pneumonia.

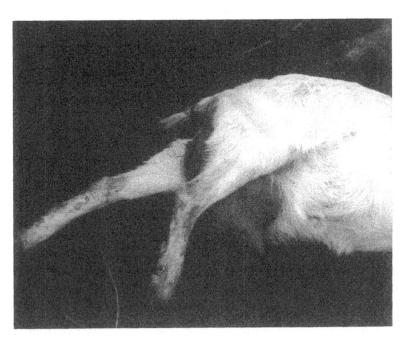

Fig. 13.   Acute toxic mastitis in a doe.

urolithiasis

- Predominantly seen in intensively housed, indoor flocks fed on concentrates.

- Precipitation of small crystals of struvite (calcium-magnesium-ammonium phosphate) leads to progressive obstruction of the urethra.

- The problem appears to be exacerbated by early castration, since the immature male urethra is narrower and more prone to blockage.

- Clinical signs include straining, non-productive attempts at urination, abdominal discomfort (colic) and reluctance to lie down. Occasionally, hematuria may be seen. In goats, vocalization is common.

- Untreated cases will progress and develop a rupture of either the urethra or bladder. Urethral ruptures lead to extravasation of urine into subcutaneous tissues around the penis ("water belly"). Bladder ruptures may not be obvious until the animal develops uremia.

- Diagnosis is based upon dietary history and clinical signs.

- Treatment depends on the severity of the problem. Mild, uncomplicated cases, with partial obstruction, should be treated with tranquilizers and antispasmodics (e.g., diazepam, 0.1–0.5 mg/kg IV).[8]

- More severe cases may require amputation of the urethral process, urethrostomy or tube cystostomy.[92]

- Prevention depends upon careful control of the diet. A commercial small ruminant diet, with a calcium-to-phosphorus ratio of 2:1 is optimal. Urinary acidifiers, such as ammonium chloride (40 mg/kg/day, in feed) or ascorbic acid (3 mg/kg/day, SQ), may be used to inhibit struvite formation. Salt blocks and free access to water are also important.[92]

Neurological Diseases

caprine arthritis-encephalitis

- Extremely common in the United States as compared with Europe. An eradication program is being coordinated by the USDA.

- The causative agent, caprine arthritis-encephalitis virus (CAEV) is a slow virus (lentivirus) of the same family as those causing HIV in humans. However, there is no evidence that CAEV can be transmitted to humans.

- Both horizontal and vertical transmission are important. The virus is present in secretions, particularly colostrum and milk, but also in urogenital secretions at or around the time of kidding.[93]

- Clinical signs depend upon the age of the animal, but include arthritis, encephalitis, pneumonia, and mastitis.

- Encephalitis is most often seen in young kids (1 to 5 months old). Clinical signs include ataxia, hindlimb paresis and, ultimately, tetraparesis.[8]

- Older animals (usually >1 year old) are more likely to develop arthritis, most commonly affecting the carpal joints (Figure 14). Clinical signs include lameness, depression, anorexia, and loss of condition.[94]

- Indurative mastitis can cause a dramatic drop in milk yield within the herd. The milk typically appears normal.

- The interstitial pneumonitis in CAEV resembles that of maedi-visna in sheep.

- Diagnosis is based on typical clinical signs and can be confirmed by histopathology (mononuclear infiltrates in affected joint tissues, demyelination in the central nervous system, interstitial pneumonia), and positive serology.

- Effective control of CAEV depends on identification and elimination of infected animals; reactors to the serologic test should be culled, and new arrivals should be quarantined and tested before being allowed to mix with the rest of the flock.

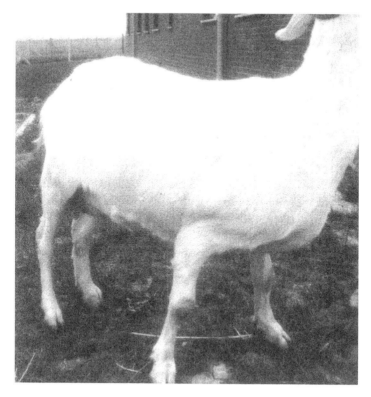

Fig. 14. Typical signs of carpal arthritis ("big knee") in a goat with CAEV.

scrapie

- Reportable disease.

- Scrapie is a transmissible spongiform encephalopathy (TSE), believed to be caused by a prion agent of the same family as those responsible for Kuru and Creutzfeld-Jacob disease (CJD) in humans, and bovine spongiform encephalopathy (BSE) in cattle.[95]

- Both sheep and goats are susceptible, although the natural disease appears to be rare in goats.

- Scrapie is relatively uncommon in the United States as compared with the United Kingdom.

- The disease has a long incubation period and clinical scrapie has been seen only in adult animals (>2 years old).

- Clinical signs include insidious weight loss, pruritus, alopecia, and nervous signs.[96]

- Treatment is futile and animals should be euthanized as soon as possible. If left untreated, animals develop progressive paralysis and die of respiratory failure.

- Diagnosis is based upon physical examination and exclusion of other common causes of these clinical signs. At present, confirmation of a diagnosis of scrapie can only be made after histopathological examination of the brain. Pathognomonic findings include vacuolation of the grey matter.

- Although no direct evidence exists for transmission of scrapie to humans, animals with clinical signs should be handled with care. Carcasses should be incinerated.

- Voluntary flock accreditation is ongoing in the United States. A genetic test for resistance to scrapie is being used to try to eliminate the disease from the Suffolk breed.[97]

listeriosis

See Chapter 3.

Skin Diseases

parasitic skin diseases

- Most commonly caused by external parasites, including mange mites (*Sarcoptes scabeii, Chorioptes spp.*, and *Psoroptes spp.*), lice, keds, ticks, and flies.[98,99] However, scrapie is an important differential diagnosis for itchy sheep.

- Sarcoptic mange is uncommon in sheep and goats. Lesions are extremely pruritic and preferentially affect the ears and periorbital skin. In severe cases, lesions then spread to axillary and inguinal regions, udder and perineal regions. *Sarcoptes scabeii* has the potential to spread to humans.[100]

- *Psoroptes ovis* (the causative agent in "sheep scab") has been eradicated from the United States, but disease due to *Psoroptes cuniculi* (a primary pathogen of rabbits) is seen in goats. Lesions are usually restricted to the ears, and cause head shaking.[8]

- Chorioptic mange is usually seen predominantly on the lower limbs and ventral surfaces of the sternum, scrotum and udder.

- Biting and sucking lice can be a problem in housed sheep and goats, particularly during the winter months. Pruritus is the predominant clinical sign.

- Problems with flies and, in particular, with fly larvae are seen in the summer. Fly strike, in which fly larvae invade the fleece and underlying skin, is a serious concern in range animals.

- Ticks are seen predominantly in animals at pasture, mainly in the summer and early fall.

- Although some of these parasites can be identified by careful visual inspection, deep skin scrapes and occasionally skin biopsies may be required to make a definitive diagnosis of mange infestations.

- A variety of topical parasiticides, including malathion, permethrin, and coumaphos, have been recommended for the treatment and control of external parasites of small ruminants. These agents are available in a variety of forms, including sprays, dips, pour-on liquids, and medicated ear tags.[101]

orf

See Chapter 3.

caseous lymphadenitis

- Causative agent is *Corynebacterium pseudotuberculosis*. Both sheep and goats are susceptible to infection and clinical disease.[102]

- The predominant clinical sign of caseous lymphadenitis is abscessation of the external lymph nodes, particularly those of the head and neck.

- Internal lymph nodes (mediastinal and mesenteric) may also become involved (Figure 15). In these cases, animals present with signs of chronic disease, including emaciation and respiratory problems.

- If untreated, the abscesses grow and finally rupture, releasing thick, green-white pus which is full of bacteria.

- The main route of transmission appears to be through ingestion of the organism, although the organism may also spread on the clippers used to shear animals with abscesses.[42]

- Diagnosis of *C. pseudotuberculosis* infection is confirmed by isolation of the organism from fresh abscesses.

- Although drainage or excision of the abscess has been advocated, control of *C. pseudotuberculosis* depends on the successful elimination of carrier animals from the herd.

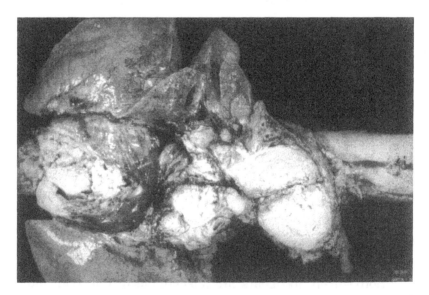

Fig. 15. Internal lymph node abscessation in a buck with caseous lymphadenitis.

woolslip

- Woolslip is a non-pruritic disease, usually presenting as simple hair loss (alopecia). The underlying skin is intact and is not irritated.

- Sporadic cases are seen after winter shearing.

- Although the precise cause of woolslip is unknown, the hair follicle itself is still intact. No treatment is required, and the fleece will grow back.

Nutritional/Metabolic Diseases

metabolic diseases

- As with the large ruminants, the most important metabolic diseases in sheep and goats are hypocalcemia, pregnancy toxemia and hypomagnesemia (Table 10).

- Clinical signs are often highly confusing, and a broad-based approach to treatment is essential if animals are to be saved. Quite often, the diagnosis is only confirmed by a successful response to therapy.

- Pregnancy toxemia is typically seen in late pregnancy. Affected animals are usually either (a) overweight or (b) thin. Treatment is relatively easy in sheep, but difficult in goats.

- Hypocalcemia is rare and usually mild in goats; cases respond well to intravenous calcium.[8]

- Hypomagnesemia is relatively uncommon in goats.

selenium/vitamin E deficiency ("white muscle disease")

- Deficiency of selenium and/or vitamin E appears to predispose lambs to white muscle disease.

- Clinically, lambs appear stiff, have difficulty rising and grow poorly.

- At necropsy, affected muscles are pale and show evidence of necrosis. The disease predominantly affects skeletal muscle, although cardiac muscle may also be involved.

- Treatment of stiff lambs with injectable vitamin E/selenium compounds can be effective.[42]

- In a laboratory setting, the disease should be rare with animals on commercial rations. However, in outdoor flocks on selenium-deficient soils, white muscle disease can be a significant problem.

- Prevention involves prophylactic supplementation of the ewe (either in-feed or by injection) before breeding and lambing.[103]

TABLE 10. METABOLIC DISEASES IN SMALL RUMINANTS

	Pregnancy Toxemia	Hypocalcemia	Hypomagnesemia
Clinical signs	Separated from flock; fine tremor; blind.	Ataxia, depression, recumbency, bloat, constipation.	Excitable, nervous, stiff gait, twitching, convulsions, death.
History	Last month of pregnancy; old, fat or broken-mouthed ewes; usually twin fetuses.	Typically late pregnancy.	Usually seen in sheep grazing lush grass in spring or fall. Also seen in ewes post-lambing.
Diagnosis	Urine tests positive for ketones.	Clinical signs and response to therapy.	Clinical signs and response to therapy.
Immediate treatment	50–100 ml glucose SQ; 50 ml propylene glycol PO.	20-40 ml CaMP IV; 20-40 ml 20% glucose IV. 50-100 ml CaMPD SQ. 150 ml 20% glucose PO.	
Continued treatment	Repeat BID.	Monitor response.	Monitor response.
Prevention	Identify and separate out ewes with twin fetuses; feed increasing plane of nutrition in last 6 weeks of pregnancy.	Increase Ca and Vitamin D3 content of ration in last 6 weeks of pregnancy.	Increase Mg intake by supplementing ration, using mineral blocks or intraruminal Mg bullets.

Key: CaMPD, calcium-magnesium-phosphorus-dextrose solution.

general approach to the treatment and control of infectious diseases in small ruminants

Antimicrobial Therapy

Systemic antibiotics are the cornerstone of the treatment of most bacterial infections in small ruminants. There are at present no specific therapies for viral infections in small ruminants, but adjunct therapy with antibiotics may be beneficial if secondary bacterial infections develop.

The rational use of antibiotics requires a knowledge of the most likely causal organism and, if possible, laboratory evidence of sensitivity to a particular antibiotic. Some suggestions for appropriate antibiotics are given in Table 11.

TABLE 11. ANTIBIOTIC RECOMMENDATIONS FOR SMALL RUMINANTS

Generic Name	Indications	Dose	Reference(s)
Oxytetracycline	Pneumonia, chlamydiosis	6–11 mg/kg IV, IM SID	104
Benzathine penicillin G	Pneumonia	44–66,000 IU/kg SQ SID	23
Procaine penicillin G	Pneumonia, foot rot	40,000 IU/kg IM SID	104
Ampicillin	Mastitis, diarrhea	15 mg/kg SQ SID	104
Dihydrostrepto-mycin	Foot rot	20 mg/kg SID	21
Sulphadi-methoxine	Coccidiosis, metritis	110 mg/kg IV, PO SID	104
Ceftiofur	Pneumonia	2.2 mg/kg IM SID	105
Tylosin	Mycoplasmal infections	10 mg/kg SQ SID	23

It should be noted that few of these products have been approved by the Food and Drug Administration for use in sheep and/or goats. Sheep and goats are still considered relatively minor species in the United States, and it is not financially justifiable for pharmaceutical manufacturers to go through the testing required to obtain FDA approval. Nevertheless, these agents are used extensively in small ruminants, and the recommendations given in Table 11 reflect the dosages that have been found to be clinically effective in the farm and/or laboratory setting. Any concerns regarding the use of non-approved drugs in a research protocol should be addressed to the attending veterinarian.

In addition to systemic treatment with antibiotics, fluid therapy is critical in animals with signs of dehydration (e.g., neonates with diarrhea) or circulatory collapse (e.g., toxic acute mastitis). For the treatment of simple dehydration, oral electrolyte solutions and perhaps subcutaneous boluses of balanced polyionic fluids (lactated Ringer's or normal saline) would be appropriate.

If dehydration is more severe, for example as a result of severe diarrhea or an intestinal obstruction, electrolyte and acid-base disturbances are possible. With severe diarrhea, metabolic acidosis and hyponatremia may develop, while an intestinal obstruction is more likely to produce hypochloremia and hypokalemia. In these cases, intravenous fluids should be administered through indwelling catheters in either the cephalic or jugular vein. The rate of fluid administration will be determined by the degree of circulatory compromise, but should be in the range of 20–40 ml/kg/day. The choice of fluids can be critical in cases with electrolyte/pH disturbances, and prompt veterinary attention should be sought for all animals with severe dehydration. In animals with severe endotoxic shock, specific treatment to counter the effects of endotoxin (e.g., intravenous administration of flunixin meglumine at a dose rate of 1.1 mg/kg) may be lifesaving.

Localized infections of the eye can be managed with topical oxytetracycline or cloxacillin antibiotic eye ointment or drops. In more severe cases, a subconjunctival injection of a long-acting antibiotic can be used to provide sustained release of antibiotic around the eye. Prior to initiating antibiotic therapy, it is important to check the integrity of the cornea with fluorescein dye; many topical eye creams/drops contain both antibiotics and corticosteroids, and although the steroids reduce inflammation and irritation, they should be avoided in the early stages of a corneal ulcer as they retard corneal healing. Simple antibiotic creams should be used until the cornea has healed, whereupon the cream can be changed to a combination antibiotic/steroid preparation.

Localized antibiotic preparations are also very useful in the treatment of reproductive diseases. Intrauterine pessaries can be used to supplement systemic antibiotics in animals with metritis, while intramammary antibiotic preparations can be injected directly into udders affected with mastitis.[21]

Vaccination Strategies

Vaccines have been developed for a number of important bacterial and viral diseases in small ruminants (Table 12). Although developed initially for sheep, a number of these products can be used in goats.

Parasitic Diseases

The control of internal and external parasites is extremely important, particularly in sheep and goats that are maintained at pasture. Recommendations for the control of **internal parasites** are presented in Table 13, but specific programs should be developed for each facility and should, wherever possible, involve both parasite avoidance (by pasture rotation) and prophylactic anthelmintic treatment.[84]

External parasites are best controlled by routine use of insecticides. These can be purchased as dips, sprays or drop-on products.[18,106] In a research setting, particular care should be taken to avoid sending animals with either fluid discharges (e.g., runny eyes, diarrhea) or open lesions (e.g., surgical incisions) out to pasture, since these animals are prone to attack by flies (myiasis or "strike"). If flies are a problem in indoor facilities, fly screens, adhesive fly paper, and/or electrical insect killers should be used to protect the animals.

anesthesia and analgesia

Procedures which have the potential to produce more than momentary pain or distress should be performed with the appropriate use of anesthetics and/or analgesics. General anesthesia is used for highly invasive or otherwise painful procedures, while local anesthesia may be appropriate for procedures in which desensitization of a small, localized anatomic site is all that is required. Analgesics should be used as an adjunct to any procedure where there is a likelihood of pain lasting beyond the duration of any anesthesia, for example after major surgery.

TABLE 12. RECOMMENDED VACCINATION STRATEGIES FOR SHEEP AND GOATS

Disease	Vaccine	Schedule
Enterotoxemia	*Clostridium perfringens* C and/or D toxoid	Ewes: 4 weeks before lambing Does: 4 weeks before kidding Lambs and kids: 8 and 12 weeks of age
Tetanus	*Clostridium tetani* toxoid	Usually combined with *C. perfringens* vaccine (see above)
Orf	Live, attenuated orf vaccine	Ewes: at least 8 weeks before lambing Does: at least 2 weeks before kidding
EAE	Inactivated *C. psittaci*	Ewes: 6 and 2 wks before breeding. Annual boosters thereafter
Campylobacter	Inactivated *C. fetus* and *C. jejuni*	Usually combined with EAE vaccine (see above)
Caseous lymphadenitis	Inactivated *C. pseudotuberculosis*	Ewes and does: 2 doses initially, then annual boosters thereafter Usually combined with *C. perfringens* vaccine (see above)
Colibacillosis	Inactivated *E.coli* vaccine	Ewes: 6 and 2 weeks before lambing Does: 6 and 2 weeks before kidding

Principles of General Anesthesia

Anesthesia should be used for any procedure which causes more than momentary pain. Although the pharmacology of anesthetic agents is beyond the scope of this book (interested readers should refer to texts on veterinary anesthesia),[110-112] dose rates for the most commonly used anesthetic agents are provided in Table 14. The minimum dose of anesthetic agent required to provide an adequate depth of anesthesia should be used; since there can be tremendous variation between individual animals

TABLE 13. ANTHELMINTICS FOR SMALL RUMINANTS

Chemical Name	Route	Activity				Dose (mg/kg)	Reference(s)
		GI	Lung	Tape	Fluke		
Albendazole	PO	+[a]	+	+	+	5 (7.5 for fluke)	107
Mebendazole	PO	+	+	+	–	15	21
Fenbendazole	PO	+	+	+	–	5	108
Thiabendazole	PO	+	+	–	–	44	108
Levamisole	PO, SC	+[a]	+	–	–	7.5	109
Ivermectin	PO, SC	+[a]	+	+	–	0.2	80

Key: + active against all larval and adult forms; – inactive against adult or larval forms; +[a] active against larval, adult, and dormant (hypobiotic) forms.

in terms of the amount of anesthetic required to produce a given level of anesthesia, care must be taken when administering these agents. Animals that have been treated with sedatives or tranquilizers before surgery may require significantly less anesthetic than animals which have not been premedicated.

Note: Whenever possible, decisions regarding the appropriate use of anesthetics and analgesics should be made in consultation with a qualified veterinarian.

Abbreviations for the route of administration of anesthetic agents are PO (oral), IV (intravenous), IM (intramuscular), and SC (subcutaneous).

Characteristics of Commonly Used Injectable Anesthetic Agents

barbiturate acid derivatives

- Include thiopental, methohexital and pentobarbital.

- Although it has been used as an anesthetic agent for many years, pentobarbital has largely been superceded by short-acting barbiturates. Pentobarbital should be used only as a means of euthanasia in small ruminants (see later).

- Methohexital and thiopental are ultra-short-acting agents that are most often used for the induction of general anesthesia. An intravenous dose of either agent will provide sufficient anesthesia to allow the introduction of an endotracheal tube.

ketamine hydrochloride

- Ketamine is a dissociative agent which is extremely popular in both clinical and research settings. It provides good analgesia but poor muscle relaxation.

- A single intravenous dose of ketamine (11 mg/kg) provides 15-30 minutes of anesthesia and can be used for castration and other minor procedures on adult sheep and goats.[8]

- If longer periods of anesthesia are required, or if muscle relaxation is important, ketamine can be combined with xylazine (see Table 14).

- Ketamine-xylazine and ketamine-detomidine combinations can be used to induce general anesthesia and facilitate the introduction of an endotracheal tube for gaseous anesthesia.

- Salivation can be a problem with xylazine. However, pretreatment of the animal with atropine or glycopyrrolate (an antisialogogue) tends to increase the viscosity of the saliva, and is best avoided.[110]

- Ketamine is under consideration for inclusion as a controlled substance by the Drug Enforcement Administration (DEA).

propofol

- Substituted phenol derivative.

- Propofol appears to be extremely safe in both sheep and goats.[118,119]

- Intravenous boluses induce rapid loss of consciousness and are suitable for minor procedures, including endotracheal intubation.

- For longer procedures, anesthesia can be maintained with inhalatory agents or with a continuous infusion of propofol.[34]

TABLE 14. ANESTHETIC DRUGS FOR SMALL RUMINANTS

Agent	Dose/Route	Duration of Effect	Reference(s)
Thiopental	10–15 mg/kg IV	20 minutes	113,114
Methohexital	4 mg/kg	5–10 minutes	110
Pentobarbital	20–30 mg/kg IV	60 minutes	8,115
Ketamine	11 mg/kg IV, IM (goats) 22 mg/kg IV	15–30 minutes 23 minutes	8,116
Ketamine *plus* Xylazine	11 mg/kg IM + 0.22 mg/kg IM	25–30 minutes	116,117
Propofol	3.5 mg/kg IV (sheep) 4 mg/kg IV (goats)	Induction	118 119
Tiletamine *plus* Zolazepam	13.2 mg/kg IV (sheep) 5.5 mg/kg (goats)	42 minutes 60 minutes	120
Alphaxalone- alphadolone	2.2 mg/kg IV (sheep) 0.23 mg/kg/minute IV infusion	10 minutes	110 121

alphaxalone-alphadolone

- Steroid derivative.

- Alphaxalone-alphadolone appears to be extremely safe in both sheep and goats, and has been used with some success in lambs and kids.[110]

- Intravenous boluses induce rapid loss of consciousness and are suitable for minor procedures, including endotracheal intubation.

- Continuous intravenous infusions of alphaxalone-alphadolone have been used to maintain anesthesia.[121]

Principles of Gaseous Anesthesia

The use of inhaled gaseous anesthetic agents allows more precise control over the depth of anesthesia and facilitates artificial ventilation of animals that have been intubated. More complete descriptions of the principles of inhalation anesthesia can be found in texts on veterinary anesthesia.[110-112]

Administration of gaseous anesthetic agents usually requires specialized equipment such as an anesthetic machine which regulates the delivery of anesthetic and oxygen. In view of the potential hazards of human exposure to gaseous anesthetic agents, an active scavenging system should be used to remove waste gases. Several commercial anesthetic machines and scavenging devices are now available for the veterinary market.

At the most basic level, gaseous anesthetic agents may be delivered through a simple face mask. This technique is only really applicable to young lambs and kids, and even these animals may require light sedation (see later) before they will accept a face mask. The mask is placed gently over the animal's muzzle and secured with a gauze tie around the animal's ears. The depth of anesthesia can be adjusted slowly by altering the setting on the precision vaporizer.

Under most circumstances in the research setting, maintenance of a patent airway in the unconscious animal is most easily accomplished by the use of an endotracheal tube which passes between the trachea and the anesthetic machine.

equipment

- Laryngoscope with 250 to 350-mm blade.

- Cuffed endotracheal tubes (typically 8 to 11 mm for adult small ruminants).

- Wire stylet.

- Water-soluble lubricant

- 10-ml syringe for inflating the cuff on the endotracheal tube.

- Gauze roll ties for (a) applying traction to the upper and lower jaws and (b) securing the endotracheal tube.

➤ Technique for Endotracheal Intubation

1. Anesthesia is induced with a short-acting injectable anesthetic agent (Table 14).

2. The sheep is placed in sternal recumbency and the head and neck extended in a straight line.

3. The laryngoscope is placed over the back of the tongue and directed towards the larynx.

4. The blade of the laryngoscope is used to apply downward pressure on the epiglottis, thereby exposing the vocal cords and the opening to the trachea.

5. The endotracheal tube, containing a wire stylet, is passed along the blade of the laryngoscope and gently inserted into the glottis (Figure 16). As soon as the tip of the tube enters the trachea, the stylet is withdrawn. The endotracheal tube is then advanced into the trachea.

6. The cuff of the tube is inflated with air and the tube is secured in place with a tie which passes behind the lower incisors and ties under the animal's chin.

7. Location of the tube within the trachea, rather than the esophagus, is confirmed by detecting passage of air through the tube as the animal breathes.

8. The external end of the endotracheal tube is connected to the anesthetic machine. A small amount of ophthalmic ointment is placed in each eye to prevent dessication.

9. At the conclusion of the surgery, the gauze tie is untied, and the cuff deflated, but the tube is left in place until the animal assumes sternal recumbency. Premature removal of the tube increases the risk of aspiration of rumenal contents and should be avoided.

Although it takes some practice, it is also possible to intubate sheep and goats without direct visualization with a laryngoscope. For this technique ("blind" intubation), the animal should be in sternal recumbency with its head and neck in full extension. The anesthetist palpates the external surface of the larynx with his/her non-dominant hand, then slides the endotracheal tube into the animal's mouth with the other hand. The larynx is then lifted slightly to meet the advancing endotracheal tube. If the tube is inserted correctly, it is possible to feel the tube pass through the non-dominant hand as it moves down the trachea. However, correct positioning should be confirmed as described previously.

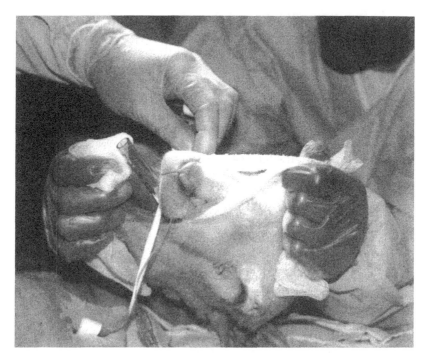

Fig. 16.   Endotracheal intubation technique.

Characteristics of Commonly Used Gaseous Anesthetic Agents

halothane

- Advantages: rapid onset of anesthesia with good muscle relaxation.

- Disadvantages: may induce hypotension and cardiorespiratory depression; sensitizes the myocardium to catecholamines.

- Minimum alveolar concentration is approximately 1% in sheep and 1.3% in goats.[122,123]

- For maintenance of general anesthesia, halothane is typically used at concentrations of 1.5–2%.

isoflurane

- Advantages: Produces effective anesthesia, with less adverse effects than either halothane or methoxyflurane.

- Disadvantages: May induce some degree of respiratory depression and hypotension.

- Minimum alveolar concentration is approximately 1.6% in sheep and 1.5% in goats.[122,123]

- Typically administered at concentrations of 1–1.5%.

methoxyflurane

- Advantages: produces effective anesthesia with good muscle relaxation.

- Disadvantages: prolonged recovery possible; potential renal and/or hepatic toxicity.

- Minimum alveolar concentration is approximately 0.26% in sheep.[123]

- Typically administered at concentrations of 0.4–1%.

nitrous oxide

- True anesthetic agent in humans (used in dentistry).

- Traditionally used to provide background analgesia in cats and dogs.

- Less popular in ruminants because the nitrous oxide readily diffuses into the rumen and may produce rumenal tympany.[110]

Principles of Local Anesthesia

Local anesthesia involves the selective desensitization of specific, defined sites, typically involving the skin and subcutaneous tissues. Four forms of local anesthesia are used in small ruminants:

Surface anesthesia involves the topical application of local anesthetic, usually onto skin, eye or mucous membrane. The most common applications for this form of anesthesia are desensitization of the eye (for ophthalmic procedures) and desensitization of the skin over catheter sites.

Local infiltration/ring blocks are most commonly used for minor procedures such as skin biopsy and castration. A line block can also be used for surgical approaches to the left flank in standing animals.[124] The anesthetic agent is injected directly into the skin and subcutaneous structures.

Both surface and local anesthesia are safe and easy to perform, requiring no specific knowledge of the innervation to a structure.

Intravenous regional anesthesia (IVRA) involves the systemic injection of local anesthetic agents into a vein which drains the area to be operated. In ruminants, IVRA is most often used for anesthetizing the foot to allow amputation of a damaged digit.[125]

Nerve and spinal blocks are extremely useful techniques which desensitize all of the structures innervated by a given nerve or nerve root. In small ruminants, the most common examples include blockade of the cornual branches of the lacrimal and infratrochlear nerves for disbudding goats,[126] and lumbosacral epidural blockade for obstetric procedures.[124]

In sheep and goats recovering from hindlimb orthopedic surgery or major abdominal procedures, effective postoperative analgesia has been reported with lumbosacral epidural injections of α-2 agonists (e.g., xylazine, detomidine) and opiates (e.g., morphine, fentanyl). The drugs can be delivered either by direct injection,[127] or through an indwelling catheter.[128]

Characteristics of Commonly Used Local Anesthetic Agents

lidocaine hydrochloride

- Direct injection of 2% lidocaine into the skin and subcutaneous structures produces excellent desensitization within 3 to 5 minutes.

- Lidocaine nerve blocks typically require 5 to 20 minutes to become fully effective.

- Lidocaine is available with or without epinephrine; the epinephrine acts as a vasoconstrictor and extends the duration of the block. Solutions containing epinephrine should not be injected into the epidural space or directly into the edges of wounds because of the risk of cord injury or skin necrosis.

- Systemic overdoses of local anesthetic can be fatal;[129] particular care should be taken to minimize the volumes of lidocaine used in disbudding procedures in young kids.[8]

Sedation and Tranquilization

Minor procedures which are not painful, but which require the animal to be relatively immobile, may be performed with the aid of sedatives or tranquilizers. Sedative agents by definition cause some degree of drowsiness and, at high doses, will cause the animal to lose consciousness. Tranquilizers exert a calming effect without loss of consciousness, even at high doses. In practical terms, both types of agent may be used to facilitate examinations and minor procedures (e.g., blood collection) on small ruminants. However, these drugs are most often used as preoperative treatments (premedicants) for animals that are scheduled for general anesthesia. Premedication with a sedative and/or tranquilizer serves both to calm the animal and to reduce the dose of anesthetic agent required to induce unconsciousness. Post-operatively, these drugs may also be used to smooth recovery from anesthesia. Recommendations on the use of sedative/tranquilizers are given in Table 15.

TABLE 15. COMMON SEDATIVES/TRANQUILIZERS FOR SMALL RUMINANTS

Agent	Dosage/Route	Reference(s)
Acepromazine	0.05–0.1 mg/kg IM	110
Xylazine	0.1–0.2 mg/kg IV (sheep)	129
	0.05–0.1 mg/kg IV (goat)	
Detomidine	30–90 µg/kg IM, IV (sheep)	130
Diazepam	0.2–0.5 mg/kg IV	131

Analgesia

Any procedure which has the potential to produce post-procedural pain should involve the use of an analgesic. The assessment of pain and distress in laboratory animals has become an important focus of research in recent years.[132] Sophisticated systems for assigning pain scores have now been developed for sheep,[132-134] and although less work has been performed with goats, similar schemes may be applied to this species. Many

IACUCs are now requiring investigators to provide documentation of post-procedural pain scores and analgesia regimens (Dr. SimonTurner, personal communication, 1997).

Pain in small ruminants usually causes dullness, lethargy and a general lack of interest in the surroundings. The animal may appear reluctant to move around the pen, and refusal of food and water is common. Vocalization is common in goats, and abnormalities of facial expression may be apparent. Animals in severe pain may have an elevated respiratory rate, with short, shallow breaths. Grinding of teeth (bruxism) is usually a sign of severe pain in small ruminants. However, it is important to note that sheep, in particular, may tolerate severe pain without overt signs of either pain or distress.

Characteristics of Commonly Used Analgesic Agents

Three classes of analgesics are commonly used in small ruminants: opiates, non-steroidal anti-inflammatory drugs, and α-2 agonists.

opiates

- A number of natural and synthetic opiates have been used in sheep and goats.

- In general, opiates agents are used for procedures where moderate to severe pain is anticipated.

- The use of morphine is becoming less common, mainly as a result of the introduction of butorphanol and buprenorphine.

- Fentanyl is proving to be very popular in small ruminants; it can be given by the intravenous, epidural, or transdermal route for post-procedural analgesia.[133,135]

- Morphine, fentanyl, buprenorphine and, most recently, butorphanol are classified as controlled substances by the DEA.

Note: Controlled substances should be kept in a lockable, double-entry safety cabinet. Written records should be maintained for each substance, in compliance with DEA.

---

non-steroidal anti-inflammatory drugs (NSAIDS)

- In many situations, much of the pain after surgery is a result of tissue damage and inflammation. Under these circumstances, non-steroidal anti-inflammatory drugs (NSAIDs) can be extremely useful.

- A number of NSAIDs have been used both clinically and experimentally in small ruminants,[136] including aspirin,[112] phenylbutazone,[133] flunixin meglumine,[137] and carprofen.[138]

a-2 agonists

- Xylazine, detomidine and medetomidine have been used in small ruminants.

- Goats are extremely sensitive to the effects of xylazine.[110]

- Xylazine may cause increased salivation; pretreatment with atropine or glycopyrollate is rarely worthwhile, since this only serves to increase the viscosity of the saliva.

- Epidural administration of these agents provides long-lasting analgesia.[139]

Information on commonly used analgesics is provided in Table 16. More detailed information can be found in veterinary anesthesia textbooks.[110-112,115]

Perianesthetic Management

The administration of sedative, tranquilizing and anesthetic agents is not without risk. It is therefore critical that animals undergoing anesthesia receive appropriate supportive care before, during and after the procedure.

Only healthy animals should be subjected to anesthesia for elective purposes. Non-elective surgery, for example for veterinary treatment, should be performed only after consultation with a qualified veterinarian.

care prior to anesthesia

A physical examination, including measurement of rectal temperature and auscultation of the heart and lungs, should be performed on all animals before any anesthetic agent is admin-

TABLE 16. ANALGESICS COMMONLY USED IN SMALL RUMINANTS

Agent	Dose/Route	Pain Relief	Duration	Reference(s)
Aspirin	100 mg/kg PO	Mild to moderate	12 hr	112
Phenylbutazone	6 mg/kg IM, IV, PO	Mild to moderate	6–12 hr	140
Flunixin meglumine	1.1–2.2 mg/kg IM, IV	Moderate to severe	NA	137
Xylazine	0.05–0.1 mg/kg IM	Moderate to severe	< 2hr	112
Butorphanol	0.05–0.1 mg/kg IM	Mild to moderate	4 hr	141
Buprenorphine	0.005 mg/kg IM	Moderate to severe	4–6 hr	112
Morphine	Up to 10 mg total dose (IM route)	Severe	2–4 hr	112
Fentanyl	10–20 µg/kg IV	Severe	2–4 hr	135

istered. Blood samples should be collected from aged or pregnant animals, as well as from any animal scheduled for a complex procedure (e.g., neurosurgery or cardiothoracic surgery), in order to rule out the possibility of underlying disease. If any of these examinations produce abnormal results, the animal should be returned to its housing and veterinary attention sought.

Assuming that the animal is healthy, it should be weighed so that the appropriate dose of anesthetic/analgesic may be determined.

For long procedures, or for procedures in which substantial fluid loss is anticipated, intravenous fluids should be administered. An intravenous catheter should be placed in either the cephalic or jugular vein prior to induction of anesthesia; once the anesthetic agent has been administered, the catheter is flushed with heparinized saline (100 IU/ml) and connected to an intravenous administration set attached to a bag of sterile fluids. Lactated Ringer's saline or normal (0.9%) saline are appropriate for most procedures. For routine maintenance of circulating blood volume, administration rates of 10–20 ml/kg/hour are generally appropriate. If more aggressive fluid therapy is required (for example if the animal is losing large volumes of blood), administration rates of up to 20–40 ml/kg/hour may be used, and colloids may be used in place of polyionic fluids.

care during anesthesia

The anesthetized animal should be monitored closely throughout the duration of the procedure to determine that the anesthetic depth is sufficient and to assure that the animal remains physiologically stable. In this regard, the following procedures should be adopted:

The **depth of anesthesia** should be monitored prior to and during surgical or other experimental manipulations. Useful parameters to monitor anesthetic depth include:

1. **Jaw tone.** Loss of resistance to opening of the mouth provides a crude estimate of anesthetic depth.

2. **Changes in heart and respiratory rates and character.** Depending upon the anesthetic regimen, increases in the heart rate or in the respiratory rate and depth may indicate an insufficient level of anesthesia. However, both heart rate and respiratory rate are also affected by changes in other physiological parameters (such as arterial oxygen and carbon dioxide levels), so care must be taken in their interpretation.

3. **Pedal reflex.** With the leg extended, the skin between the claws is pinched firmly (commonly with a hemostat). At a surgical plane of anesthesia, the animal should not withdraw the limb in response to this stimulus.

4. **Palpebral reflex.** Light brushing of the fingertips across the lower eyelid produces a blink reflex in the conscious animal. At a surgical plane of anesthesia, this reflex is absent.

5. **Eye position.** Under most anesthetic regimens, the position of the eye can be used as an indicator of anesthetic depth. Immediately after induction, when the animal is lightly anesthetized, the eye sinks in the orbit so that the pupil is partially obscured by the lower eyelid. As the depth of anesthesia increases, the eye returns to a more central position. It should be noted that under

ketamine anesthesia the eye is fixed and central, irrespective of the depth of anesthesia. If ketamine is used for induction and the animal is then maintained on inhalation agents, eye position can be used as a reliable monitoring aid only after the effects of the ketamine dissipate (typically 30 minutes after administration).

6. **Pupillary light reflexes**. The pupillary light reflex (PLR), involving constriction of the pupil in response to direct illumination, should be present in the anesthetized animal. As anesthesia deepens, the PLR becomes more sluggish. If the PLR is absent, the animal is too deep and aggressive measures (e.g., immediate cessation of anesthesia administration and delivery of large values of 100% oxygen) must be taken to lighten the plane of anesthesia before the animal dies.

The **adequacy of cardiovascular function** should be assessed by monitoring:

1. **Cardiac and respiratory rate, rhythm, and character.**

2. **Electrocardiographic (ECG) pattern.**[142]

3. **Blood gas levels and pH.** Normative values are given in Table 4.

4. **Arterial blood pressure** by direct or indirect techniques.[110]

5. **Capillary refill time (CRT)** at oral mucosa (should be less than 2 seconds).

The animal's **body temperature** should be maintained to prevent hypothermia. This can be a particular problem during procedures in which a body cavity is exposed. Strategies for preventing hypothermia include:

1. Placement of the animal on a blanket or **heated pad**, preferably containing circulating water or air.

2. Administration of **warm fluids** through the intravenous catheter.

3. Increasing the ambient temperature of the procedure/operating room.

care following anesthesia

It is critical that steps be taken to assure that the animal returns to its normal physiologic state following anesthesia. In this regard, the following measures should be taken:

*   The animal should be placed in sternal recumbency in the recovery pen. This will reduce the chance of bloat.

*   Monitoring of rectal temperature to assure that the animal does not become hypothermic. Avoid placing the recovering animal in a draft. Hypothermia is particularly dangerous for young lambs and kids, since they have a large surface area to volume ratio and readily lose body heat. Hypothermic animals can be covered with blankets, placed on heat pads, or placed at least 18–24" below an infrared heat lamp. Care should be taken to assure that animals can move away from the heat source if they become too hot.

*   Monitoring of pulse and respiratory rate and character.

*   Administration of supplemental fluids as necessary.

Potential Complications of Anesthesia in SmallRuminants

Regurgitation.   Regurgitation of rumenal contents has been widely reported as a serious and potentially life-threatening complication of ruminant anesthesia, since aspiration of the regurgitant fluid can lead to the development of pneumonia. The risk of regurgitation can be reduced by fasting small ruminants for a period of 24 to 36 hours before surgery.[143] Intraoperatively, a stomach tube can be passed into the rumen in order to drain any excessive fluid. If a stomach tube is not used, the animal should be positioned so that its nasopharynx lies below the level of the rumen; this is most easily achieved by means of tilting the operating table. If a tilting table is not an option, a rolled-up towel can be used to elevate the neck region so that any fluid passing into the nasopharynx flows directly out through the mouth. In either case, a cuffed endotracheal tube should be used in order to protect the airway from regurgitated material.

Bloat.   Gaseous bloat, otherwise known as rumenal tympany, can develop during anesthesia because the animal is unable to eructate. Under most circumstances, the degree of bloat is mild and does not require treatment. However, severe distension of the rumen with gas leads to pressure on both the diaphragm and caudal vena cava. Respiratory and circulatory compromise develop under these circumstances, and if untreated, the animal may die. Passage of a stomach tube will usually relieve gaseous bloat; if the stomach tube cannot be passed, needle decompression can be achieved by inserting a large-gauge (12G) needle through the skin and into the rumen at the point of maximum distension. Bloat can be prevented by preoperative fasting and passage of a stomach tube during surgery.

aseptic surgery

Although it may be difficult to apply the basic principles of aseptic surgery to farm animals, laboratory small ruminants scheduled for major surgical interventions (e.g., abdominal surgery, joint surgery) should be moved to a laboratory setting at least 7 days prior to surgery. Every attempt should be made to minimize the stress of transportation by keeping journey times to a minimum. Once settled into the laboratory accommodation, animals should be examined by a qualified veterinarian to ensure that they are free from signs of infectious disease. At this time, feet and teeth should be checked, and any vaccines or anthelmintics administered. Animals should be housed in clean stalls in order to minimize fecal and urinary contamination of the skin in the days before surgery.

Within the laboratory setting, every effort should be made to develop good aseptic surgical techniques for small ruminants.[144] The surgical site should be clipped, cleaned with an appropriate topical antiseptic solution (typically chlorhexidine or povidone-iodine), and isolated from surrounding tissues with barrier drapes. The surgical team should take the same precautions as would be appropriate for surgery in other laboratory animals, namely:

- The use of scrub suits, disposable surgical caps and disposable surgical masks. If aerosols are anticipated during surgery, eye protection should be used.

- Sterile surgical gowns and gloves should be worn by anyone coming into contact with either the surgical site or instruments used in the surgical site.

- Sterile drapes to protect the surgical site from adjacent, potentially contaminated tissues.

- Sterile surgical instruments should be used.

Surgery should be performed with an emphasis on gentle tissue handling, rigorous hemostasis and meticulous wound closure. Antibiotics should be used pre-, intra- and post-operatively as required.

postsurgical management

Following surgery, animals should be monitored at regular intervals, until the experiment has been completed. Particular attention should be paid to the following:

- Any change in appetite or behavior should be noted. Changes may indicate that the animal is in pain or experiencing other complications.

- The surgical incision should remain clean and well apposed. Although sterility of the wound site cannot be maintained after surgery, wounds may be kept clean by the appropriate use of dressings and/or bandages. This can be especially important in small ruminants, where fecal or urinary contamination is a real concern.

- Signs of infection at the surgical site may include redness, abnormal warmth or tenderness at the surgical site. Some clear serous discharge is not uncommon, but the presence of thick white, yellow, or green discharge is abnormal. Severe infections can produce systemic illness, with anorexia and an elevated body temperature.

- Any discharge should be cultured for bacteria and, if appropriate, antibiotic therapy initiated under the guidance of a qualified veterinarian. Topical application of antibacterial solutions (dilute chlorhexidine or povidone-iodine) will also be beneficial.

euthanasia

Animals should be euthanized in a humane and professional manner, out of the sight and sensory range of other animals. The specific method chosen should produce a quick, painless death with minimal distress to the animal. Individuals responsible for euthanasia of small ruminants should be thoroughly familiar with the prevailing legislation concerning humane techniques for euthanasia. Death should be confirmed by stethoscopic auscultation for absence of an audible heartbeat. Before the carcass is disposed, the diaphragm should be incised to ensure that there is no chance of recovery of spontaneous breathing.

Specific methods of euthanasia for sheep and goats are summarized in Table 17. A complete summary of recommendations for euthanasia can be found in the 1993 Report of the American Veterinary Medical Association Panel on Euthanasia.[145]

disposal of carcasses

The safe disposal of animal carcasses is controlled by local, state and federal regulations. Wherever possible, carcasses should be incinerated. Other options include microwave irradiation, chemical rendering, and deep burial.

TABLE 17. COMMON METHODS OF EUTHANASIA FOR SMALL RUMINANTS

Method	Route/Technique	Comments
Pentobarbital overdose	150–200 mg/kg IV	Use of pentobarbital requires DEA registration.
Carbon dioxide overdose	Inhalation of 60–70%	Compressed CO_2, not dry ice.
Exsanguination		Only under terminal anesthesia.
Captive bolt		Should be followed by pithing.
Free bullet		Extreme risk to personnel. Not recommended.

experimental methodology

This chapter is not intended to provide detailed information about specific experimental procedures, but rather, describes basic techniques that are the foundation for more complex procedures.

handling

It is critical that sheep and goats be properly restrained to ensure the safety of both the animal and the handler. The fleece or hair coat, ears, and tail should not be used to hold or restrain the animal.

Manual Restraint

Adult sheep and goats can be restrained manually for most non-invasive procedures of short duration.[146-148] Sheep are more docile than goats, though generally both species can be handled and manipulated safely if proper techniques are employed. If horns are present, they potentially represent an increased risk, but except for the aggressive ram and buck, standard handling techniques can be employed. The hooves can be quite sharp and can inflict injuries to personnel if care is not exercised when handling the animals.

For minor procedures (e.g., intramuscular injection or physical examination), minimal restraint is usually sufficient. The animal should be coaxed into a corner so that it is unable to escape from the handler, then gently restrained by placing one arm under the animal's neck and another on its rump.

If more secure restraint is required (e.g., for intravenous injection or oral dosing), the sheep should be placed either into lateral recumbency or onto its rump. To place the animal in right lateral recumbency, the sheep is approached from the left side. The left arm is extended under the animal's neck, and the right forelimb grasped at the level of the carpus. The right arm is placed under the animal's belly, and the right hindlimb grasped at the hock. Gentle, but steady traction on the legs will roll the animal into right lateral recumbency; the head should be secured by a second handler in order to protect it from injury.

An alternative technique for immobilizing sheep involves standing them up on their hindquarters. The animal is approached from behind, and the muzzle grasped gently with the left hand. The head is turned and lifted simultaneously, and the animal's forequarters are supported by the right hand, placed under the sternum. The sheep is then pulled backwards towards the handler, and its weight supported by the handler's legs. With the head restrained, most sheep will become passive, allowing the handler to trim feet and perform minor procedures, such as venous blood sampling.

Mechanical Restraint

In addition to manual restraint, animals can be placed in mechanical restraint devices, such as metabolism cages, slings, or even deck chairs. Unless specifically required by a project, and approved by the local IACUC, these devices should be used for only short-term restraint, not permanent housing. Slings have been used with some success as a means of preventing premature weight bearing in animals that have undergone orthopedic procedures,[149] and as short-term restraints during infusions.[150] Cradles are useful as a means of restraining conscious sheep while they undergo procedures such as abdominal ultrasound or sampling from a rumenal fistula.[151] Metabolism cages are widely used in nutritional and pharmacokinetic studies, where 24-hour fecal and urine collections are required, but they can also be used as a means of restricting the movement of animals that are hooked up to catheters, leads, or telemetry equipment.[14] Animals that are to be placed in metabolism cages should be acclimated to this restrictive environment by placing the animal in the cage for one hour or less initially, then gradually increasing the time spent in the metabolic cage. It is important that sheep or goats in metabolism cages are able to see and hear one another.

Chemical Restraint

Although sheep and goats can be trained to allow investigators to collect samples of blood, most procedures on sheep and goats should be performed with the animals either sedated or anesthetized. Recommendations for the use of chemical agents were presented in Chapter 4.

sampling techniques

Sheep and goats are used frequently in studies in which sampling of body fluids and secretions is required. Common methods for obtaining such samples are presented below.

Venous Blood

Frequently, large volumes of blood are required of sheep and goats, particularly those used for polyclonal antibody production. General principles for blood sampling include:

Volume of Sample. If too much blood is collected at a single time, the animal may develop hypovolemic shock. If blood sampling is too frequent, anemia is possible. The 1–3–6 rule can be used to estimate the amount of blood that can safely be withdrawn: 1% of the animal's body weight (equivalent to 10 ml per kg body weight) can be withdrawn every 4 to 6 weeks; 3% is the amount that can be expected to be harvested at euthanasia (exsanguination), and 6% is the total blood volume of the animal. For frequent collections, it is best to alternate between sampling sites to minimize the risk of venous thrombosis.

> **Note:** Exsanguination is only to be performed when the animal is fully anesthetized.

Sampling Vials. Samples for evaluating whole blood are usually collected into vials containing an anticoagulant, such as ethylenediaminetetraacetic acid (EDTA). Samples for the collection of plasma are collected into vials containing heparin, citrate or potassium oxalate, depending on the experimental use. Samples for the collection of serum are collected into vials containing no anticoagulant. Blood samples for serum or plasma are centrifuged at 800 to 2000 g for approximately 10 to 15 minutes, and the liquid fraction harvested as the sample. Alternatively, the liquid fraction can generally be separated from the cellular fraction if the sample is left at room temperature for approximately 30 minutes.

Sampling Devices. Disposable syringes and needles, or disposable Vacutainers®, should be used. The bore of the needle should ideally be slightly smaller than the diameter of the vessel, in order to allow rapid blood withdrawal. In lambs and kids, 20–25G needles are useful. In mature animals, 16–20G needles are preferred. After use, they should be placed in an appropriate, puncture-resistant container designated for sharp objects. To minimize the risk of needle-stick injury, the syringe and attached needle should be disposed without attempting to re-sheath or bend the needle.

Sites. Several sites can be used for obtaining venous blood samples, including the cephalic vein and the jugular vein. For terminal blood sampling, direct cardiac puncture may be appropriate.

The cephalic vein is easily accessible, being located on the cranial surface of the lower forelimb, just distal to the elbow joint. Although it is relatively easy to identify the vessel, it is quite mobile and this can make it difficult to obtain large volumes of blood from this site. Two people are generally required to obtain blood from this site; the first to restrain the animal and raise the vein, the second to collect the sample.

The jugular vein is also easily accessible and readily provides large volumes of blood. The vein is located within the jugular furrow and, although it is readily palpable, it is best visualized after clipping the hair or fleece from the neck region (Figure 17). In docile animals, the procedure can be performed with minimal restraint. If the animal is aggressive, secure restraint (with the animal sitting on its hindquarters) should be used; this will protect both the animal and the handler from accidental injury during the venipuncture procedure.

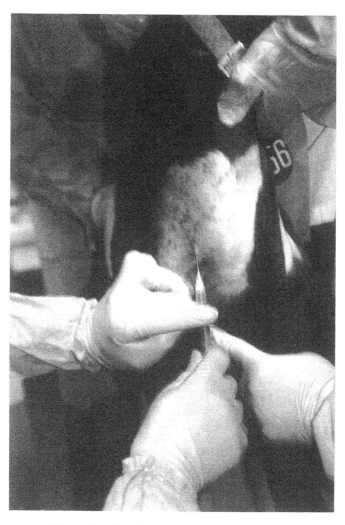

Fig. 17. Right jugular venipuncture.

Cardiac puncture should not be used in sheep or goats, except as a terminal procedure under general anesthesia. With the animal anesthetized and lying in right lateral recumbency, a long (1½ to 2 inch), 16-18G needle (attached to a 20-ml syringe) is inserted through the thoracic wall, between the ribs at the level of the point of the elbow. The needle is directed medially and slightly cranially. As the needle is advanced towards the heart, gentle suction is applied with the syringe. As soon as the needle enters the heart, blood will appear in the hub of the needle. At this point, the needle is steadied and steady backward pressure applied to the plunger so that blood flows into the syringe.

➤ *Technique for Percutaneous Venipuncture*

1. The animal must be adequately restrained.

2. Accurate venipuncture requires (a) an understanding of the anatomy and (b) practice. The location of the vein should be identified by palpation and visual inspection. The area is then clipped and swabbed with 70% alcohol to minimize the risk of introducing skin-associated bacteria into the bloodstream.

3. The vein is distended by applying pressure (with either a finger or with a tourniquet) between the work site and the heart. For the cephalic vein, a tourniquet can be placed around the proximal forearm, just below the elbow joint. This will restrict the passage of blood proximally and produce distension of the cephalic vein. The same result can be achieved by having an assistant encircle the forearm with their hand. For the jugular vein, gentle pressure on the vein at the thoracic inlet will produce distension of the vein.

4. The plunger of the syringe is pulled back slightly, breaking the air lock and allowing easier passage of the plunger through the syringe.

5. The needle is directed into the vessel at a slight angle, with its beveled edge up. The plunger of the syringe is gently and slowly pulled back as blood fills the body of the syringe. If the plunger is pulled back too quickly, the vessel may collapse and blood flow will cease.

6. Gentle manipulation, such as slight changes in the depth and angulation of the needles may improve collection if blood flow slows or ceases. Alternatively, hematoma formation around the vessel, or clotting of blood within the needle itself, may necessitate changing withdrawal sites and/or replacement of the needle.

7. Once the sample has been collected, the needle should be withdrawn and firm pressure applied to the puncture site in order to minimize hematoma formation.

vascular catheterization

Although direct collection through a hypodermic needle is feasible, repeated blood sampling from a single site is more readily achieved by means of an indwelling vascular catheter. Techniques have been described for catheterization of a number of sites, including the jugular vein, cephalic vein, and femoral vein.

The use of long-term indwelling intravenous catheters has been associated with an increased risk of valvular endocarditis in sheep.[152] Aseptic precautions, including clipping and scrubbing of the skin in the jugular groove, should be used if catheters are to be placed.

Subcutaneous vascular access ports allow chronic vascular access without an exposed catheter.[153]

Arterial Blood

Samples of arterial blood (e.g., for arterial blood gas analysis) can be collected from either an exteriorized carotid loop[34] or the femoral artery.[154] The femoral artery is palpable just dorsal to the pectineus muscle in the groin. Meticulous preparation of the skin in this region is required because of the accumulated sebaceous and, in sheep, wax secretions. When the blood has been collected, firm pressure must be applied to the puncture site, since the risk of hematoma formation is much greater after arterial sampling.

Urine

The method of urine collection depends on (a) the frequency at which samples are required and (b) the use of the intended specimen. For most purposes, urine should be collected in a clean, dry container and stored under refrigeration if the sample is not to be used within a few hours.

If sample contamination is not a significant concern, sheep can be stimulated to urinate by gently covering their nostrils with a hand. For repeated collections, or for 24-hour urine specimens, metabolism cages or disposable diapers are preferred. Recently, the use of a neoprene body stocking has been described in lambs.[155]

Relatively clean urine samples can be collected by **catheterization** of the urinary bladder. To perform this technique, the animal should be sedated lightly (xylazine is useful) and placed in lateral recumbency. A lubricated, sterile rubber catheter is passed through the urethra and into the bladder; a speculum should be used to avoid inadvertent catheterization of the suburethral diverticulum. Once the catheter has entered the bladder, urine should flow through the catheter under gentle aspiration from an attached 20-ml syringe.

If extremely clean urine specimens are required, for example in the diagnosis of urinary tract infections, **cystocentesis** can be used in young lambs and kids. The animal is sedated and then positioned in dorsal recumbency. The caudal ventral abdomen is clipped and prepared as for aseptic surgery. The urinary bladder is identified by gentle palpation and immobilized against the ventral abdominal wall. A 20–22G hypodermic needle is passed through the skin and into the bladder. Urine is then aspirated by gentle suction from an attached syringe.

Cerebrospinal Fluid

Samples of cerebrospinal fluid may be required for both experimental and diagnostic purposes. The collection of cerebrospinal fluid (CSF) usually involves:

- Anesthesia of the animal, since significant damage to the nervous system can occur if the animal is not completely immobile during sampling.

- Aseptic techniques, including aseptic preparation of the sampling site.

- Insertion of catheters or spinal needles.

- The most common sites for CSF collection are the cisterna magna and the lumbosacral epidural space.

Cisterna magna. Single specimens of CSF can be collected by direct needle centesis of the cisterna magna in the anesthetized animal. If repeat samples are required, catheterization of the cisterna magna is recommended;[156] the advantages of this procedure are that (a) CSF can be sampled in a conscious animal and that (b) the CSF samples are generally free of blood contamination. Chronic CSF sampling may be performed through a catheter which is anchored to the top of the skull with dental acrylic or cyanoacrylate glue.[157]

Lumbosacral epidural space. This site can be used to sample CSF through an indwelling catheter which is tunneled subcutaneously to exit externally through the skin.[158]

➤ *Technique for Lumbosacral CSF Sampling*

1. Catheterization is performed with the anesthetized animal in sternal recumbency.

2. An imaginary line is drawn between the cranial borders of the wings of the ilium; the lumbosacral space is located at the intersection of this line and the animal's dorsal midline, at the junction of the sixth (or seventh) lumbar and first sacral vertebrae.

3. The skin in this region is clipped and prepared for aseptic surgery. An 18G, 3.5" spinal needle is inserted into the lumbosacral space, at a right angle to the skin. A distinct "popping" sensation is felt when the needle penetrates the interarcuate ligament.[124] The position of the tip of the needle within the epidural space should be confirmed by the identification of CSF in the needle hub. The flow of CSF is encouraged by applying gentle suction to the needle with a 3-ml syringe, or by manual occlusion of the jugular vein at the level of the thoracic inlet.

Synovial Fluid

The shoulder, elbow, stifle and hock joints are easily accessible in small ruminants. In view of the risk of introducing infection into the joint, all synovial fluid aspirations should be performed using aseptic surgical techniques. The skin over the joint should be clipped and prepared with topical antiseptic, and the investigator should wear sterile disposable surgical gloves. The anatomical landmarks for synovial fluid collection have been described in goats,[8] and are similar in sheep. A 3- or 5-ml syringe and a 21G needle are appropriate for most animals.

Normative values for synovial fluid in goats have been reported.[8]

Bone Marrow

Sampling of bone marrow from sheep and goats should be performed with the animal under general anesthesia. Full aseptic technique, including preparation of the surgical site and the surgical team, is mandatory. Sampling sites include the proximal tibia, humerus, femur, and iliac crest.

Rumenal Fluid

The collection of rumenal fluid is an important element of many studies on ruminant physiology and nutrition. A number of techniques have been described for collecting samples; the choice of technique is determined by the frequency at which the samples are required.

> For **single measurements**, a stomach tube passed either through the mouth (in an anesthetized animal) or through the nasal cavity (in a conscious animal) can be used. The length of the tube can be estimated by measuring from the nostrils to the last rib.

> For **repeated sampling**, a temporary or permanent rumenal fistula can be useful.[159]

Feces

Fecal samples are required for both routine monitoring of parasite counts and for diagnostic purposes in animals presenting with diarrhea. Under certain circumstances, fecal samples may also be useful in research projects on ruminant nutrition, physiology and microbiology.

In most instances, small amounts of fecal material can be collected directly from the rectum, using a gloved hand. If chronic (e.g., 24-hour) collections are required, disposable diapers or metabolism cages can be used; the latter technique is more popular because it allows separation of feces from urine.

Bone

Sheep and goats are popular as *in vivo* models for orthopedic research. This has led to the development of techniques for assessing skeletal responses to surgical and medical therapies. Techniques have now been developed for minimally invasive sampling of cortical and cancellous bone from goats, by means of metallic chambers implanted in the proximal tibia.[160] Using this chamber, it is possible to collect bone samples at monthly intervals for periods of several months, without apparent adverse effects on the animal.

Milk

The collection of milk from ewes and does is common practice on most commercial sheep and goat farms, but is less common in the laboratory. With the exception of specific experimental studies looking at mammary gland function, the only real use for milk samples has, until recently, been as a diagnostic aid in detecting mastitis.

If small amounts of milk are required, hand-milking is quick and easy. The teats should be cleaned with a dry cloth in order to minimize bacterial contamination. The neck of the teat is then gripped gently between the thumb and middle and index fingers. Gentle downward pressure is then applied along the length of the teat, squeezing milk from the neck of the teat down to its tip. For routine analysis, milk can be collected in a small strip cup.

Recent developments in genetic engineering have led to the production of transgenic animals carrying human genes. Transgenic sheep carrying the genes for human insulin and clotting factors have now been developed; these animals secrete these proteins in their milk, making it relatively straightforward to isolate and purify the proteins in large quantities.[161] For these applications, commercial semi-automated milking machines are preferred, since the volumes of milk that are required are significant.

administration of medicines and compounds

A variety of routes exist for administration of both test compounds and medications to small ruminants. As with sampling techniques, it is important that the animal is securely restrained before any attempt is made to administer the compound. It is rarely necessary to sedate sheep or goats before administering compounds. Routes of compound administration include:

Oral

There are several methods for oral dosing of small ruminants, including:

1. **Incorporation of compound into the drinking water or food.** Although simple, this technique produces imprecise dosing. While this may be adequate for routine treatment of animals with mineral supplements or prophylactic antibiotics, it is unacceptable for most of the compounds that are administered in experimental studies. The technique is also unsuitable for treating sick animals, as the dose of any drug that a particular animal receives will be determined by its food and water intake, which may be depressed as a result of the disease that is being treated.

2. **Administration through a syringe or dosing gun.** This technique is popular with farmers in commercial sheep and goat enterprises because it is inexpensive and fast, but it can also be used in a laboratory setting. The test compound should be administered slowly to allow the animal time to swallow, and to minimize the risk of accidental aspiration of the liquid. Oral dosing, even with a syringe, is rather imprecise, since there is always some degree of unavoidable spillage. It cannot be recommended for studies in which precise volumes of test agents are to be given.

3. **Oral gavage** is the method of choice for delivering specific volumes of liquid test agents by the oral route. Although gavage is used in conscious neonatal lambs and kids as a means of supplying colostrum, it should

only be used in mature animals if they have been sedated or anesthetized.

4. **Capsule administration.** Oral capsules have been used with some success as a means of delivering anthelmintics and mineral supplements to small ruminants.

➤ *Technique for Oral Gavage*

1. The total length of tube to be inserted is estimated by measuring from the mouth to the last rib. This length should be marked on the tube before it is inserted into the animal's mouth.

2. If the animal has been sedated or anesthetized, a gag is placed between the molar teeth on one side to prevent the animal from chewing the tube.

3. The end of the tube is lubricated, placed in the animal's mouth and gently pushed towards the throat. In the conscious (but sedated) animal, the presence of the tube in the throat should induce a swallowing (gag) reflex; when the animal swallows, the tube is pushed gently into the esophagus and advanced slowly towards the stomach. In the anesthetized animal, the entrance to the trachea should be obscured by the presence of an endotracheal tube, so the gavage tube will pass relatively easily into the esophagus.

4. The location of the tube in the stomach is confirmed by ensuring that there is no air passing through the tube as the animal breathes. If the tube is in the rumen, it is generally possible to smell rumen gases at the end of the tube.

5. The compound is administered slowly by means of a syringe attached to the stomach tube. Volumes as large as 10 ml/kg can be administered safely in this way. It is advisable to flush the tube with a small amount of water before removal, as this ensures complete delivery of the compound and minimizes the chances of residual test compound entering the animal's airway as the tube is withdrawn.

Parenteral

The majority of veterinary medications and anesthetics are administered by the parenteral route. Intravascular, intramuscular, and subcutaneous routes are used commonly, while intradermal injections are used less frequently.

intravascular

Intravascular administration of compounds results in rapid delivery to target tissues. Unless specifically required by an experimental protocol, substances given by the intravascular route should be injected slowly, so that the animal can be monitored for evidence of adverse reactions (e.g., anaphylaxis).

Common sites for intravascular injection include the cephalic vein (Figure 18) and the jugular vein (see previous section for details).

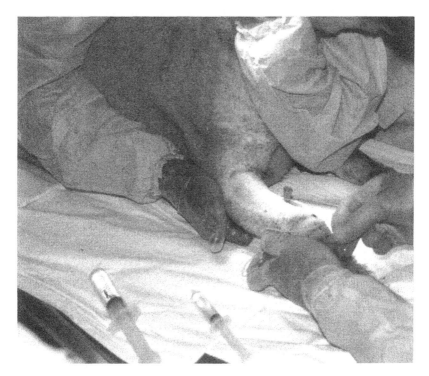

Fig. 18.   Right cephalic venipuncture using a butterfly catheter.

➤ *Technique for Intravenous Administration*

1. Liquid compounds may be administered by needle and syringe, using essentially the same approach as for blood sampling from these sites.

2. An 18–21G sterile hypodermic needle should be used. The syringe that is used should be as small as possible in order to facilitate manipulation of the syringe during injection.

3. It is important to keep the syringe and needle still during injection; if the needle comes out of the vessel, the compound will be injected into perivascular tissues. If perivascular leakage does occur, the perivascular area should be injected with sterile saline in order to dilute out the test compound and minimize the risk of tissue necrosis.

4. The compound is administered slowly and the animal monitored for evidence of respiratory or cardiovascular distress. If any adverse reactions develop, the injection should be halted and the animal examined by a qualified veterinarian.

5. Following injection, the needle is removed and bleeding controlled by slight pressure.

Chronic intravascular administration of compounds may be performed by means of implanted catheters. The jugular vein is a common site for indwelling intravenous catheters. Continuous infusions in mobile, unrestrained small ruminants can be performed using a swivel-tether or jacket system.

intramuscular

Common sites for intramuscular injection include the quadriceps muscles on the cranial aspect of the thigh (Figure 19) and the muscles on the lateral surface of the neck.

➤ *Technique for Intramuscular Injection*

1. Depending on the viscosity of the compound to be injected, an 18–21G needle should be used.

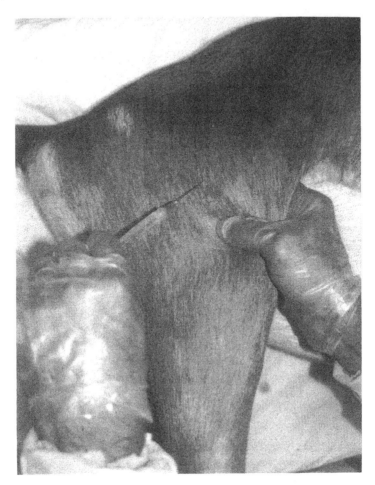

Fig. 19. Intramuscular injections into the quadriceps muscle should be performed cranial to the femur in order to avoid accidental injury to the sciatic nerve.

2. The muscle to be injected should be identified and immobilized with one hand, while the other hand manipulates the syringe.

3. The needle is inserted, bevel up, through the skin and into the muscle.

4. The syringe plunger is pulled back gently to check that the needle has not penetrated a blood vessel. If blood appears in the needle hub, the needle should be withdrawn slightly and redirected into the muscle.

5. In general, volumes of no greater than 2 to 4 ml should be injected into a single IM site.

Note: Tapping the site prior to injection may reduce "jump" when the needle is inserted.

---

subcutaneous

Common sites for subcutaneous injection include the dorsum of the neck and back and the inguinal skin fold, just cranial to the hindlimb.

➤ Technique for Subcutaneous Injection

1. The smallest bore needle is used, depending on the viscosity of the compound to be injected. An 18–21G needle is generally suitable.

2. A fold of skin is raised at the injection site.

3. The needle, with attached syringe, is inserted at a right angle to the skin fold. Care must be taken to ensure that the needle does not simply pass through the skin fold to exit the other side.

4. Relatively large volumes of fluid can be administered into each site. Avoid excessive volumes of fluids, which distend the skin and can be painful.

5. For polyclonal antibody production, antigens are usually injected in volumes of 0.1 to 0.5 ml per site.[162]

Although chronic administration of agents by the subcutaneous route may be achieved by repetitive subcutaneous injection, innovations in drug delivery have produced new techniques for compound administration, including:

Implantable osmotic pumps. Osmotic pumps are small, self-contained devices which are designed to deliver substances at a specified rate under the force of osmotic pressure. The pumps are implanted surgically into either subcutaneous sites or into the abdominal cavity.[163]

Slow-release implantable pellets. Controlled release pellets are now available for a number of biologically active agents, including antibiotics, anthelmintics, anti-inflammatory drugs, and hormones. The active ingredient is formulated in a biologically inert carrier, which degrades in a time-dependent fashion, releasing the active agent in a controlled fashion.[164]

The major disadvantage to these techniques is that they are expensive. Osmotic pumps are implanted under general anesthesia, and there is always a slight risk of anesthetic complications or post-operative wound infections. The slow-release pellets are small enough to be implanted with a custom-made injector (trocar), and this procedure may be performed in the conscious animal.

Despite these disadvantages, pumps and pellets are gaining in popularity as an alternative to daily subcutaneous injections. They eliminate the requirement for daily injections, minimizing the stress on the animals and reducing the workload for researchers and caretakers. Since the pellets and pumps release the active agent slowly throughout the day, there are none of the problems of peaks and troughs in serum concentrations which are seen after traditional subcutaneous injection.

Intraperitoneal

The intraperitoneal route is rarely used in small ruminants, mainly because of the risk of inadvertent penetration of abdominal viscera. Possible exceptions might include the intraperitoneal administration of glucose solution to neonatal lambs and kids. The preferred site for intraperitoneal injections is the caudal right abdominal quadrant.

➤ Technique for Intraperitoneal Injection

1. A long needle (typically 1 inch) is required. The smallest bore needle is used, depending on the viscosity of the compound to be injected. An 18–21G needle is generally suitable.

2. The lamb or kid is placed in dorsal recumbency, with its hindquarters raised at an angle of approximately 30 to 45 degrees to the horizontal; this will allow abdominal organs to fall away from the injection site.

3. After disinfecting the injection site with alcohol, the needle is inserted into the lower right abdominal quadrant, at an angle of 45 degrees to the body wall.

4. The plunger is withdrawn slightly to ensure that the needle has not penetrated a blood vessel or abdominal organ. Aspiration of yellow fluid suggests inadvertent perforation of the bladder, while green-brown fluid is more suggestive of entry into the intestinal tract. In either case, the needle should be withdrawn and the test compound discarded.

measurement of body temperature

The measurement of body temperature is a standard technique for assessing the health of small ruminants. The rectal temperature can be obtained with a glass clinical thermometer (apply lubricant first); the thermometer should be inserted at least 5 cm into the rectum and the temperature recorded after incubation for at least one minute. Alternatively, body temperature can be measured at the tympanic membrane using an infrared tympanic thermometer.[165] This technique is fast and easy to perform, and is extremely useful in neonatal animals, but is more sensitive to proper technique. The normal body temperature in sheep and goats, as measured by rectal thermometer, is 38.0 to 39.5°C (100.4 to 103.1°F).

polyclonal antibody production

Sheep and goats are widely used in the production of polyclonal antisera to human and animal proteins. Their popularity is in large part due to their size, since the average 70 kg sheep provides plentiful supplies of antisera. Recommendations for immunization and bleeding schedules for polyclonal antibody production in small ruminants are presented elsewhere.[13,162]

evaluation of medical devices

Sheep and goats are widely used as *in vivo* models for the evaluation of implantable medical devices, including orthopedic implants,[149,166,167] artificial heart valves and assist devices,[168,169] and vascular stents.[170,171] There is also an increasing trend towards the use of small ruminants in the preclinical evaluation of new medical and surgical techniques, including minimally invasive laparoscopic, fetoscopic, and spinal surgery.[172-174]

necropsy

Post-mortem examinations of organs and tissues are an integral part of many studies and are vitally important in the diagnosis of disease in sheep and goats.

The basic equipment that is required for necropsies on small ruminants includes:

1. Latex or rubber gloves, arm-length disposable plastic gloves, laboratory coat or coveralls, protective plastic apron, eye protection, and waterproof boots.

2. Small metric ruler.

3. Toothed and serrated tissue forceps.

4. Scalpel blades (#10 and #21) and handles (#3 and #4).

5. Dissecting and small operating scissors.

6. Metal probe.

7. Bone-cutting forceps and/or electric necropsy saw.

8. Sterile swab for bacteriological culture of tissues and exudates.

9. Blood vials (plain, EDTA) for collecting samples of blood and tissue fluid.

10. Glass or plastic vials for tissue specimens.

11. 10% neutral buffered formalin for preserving tissue specimens.

12. Syringes (3- and 10-ml) with both large bore (16–18G) and small bore (20–22G) hypodermic needles.

Additional equipment may be useful and can be added to this basic kit.

The necropsy should be performed in a designated necropsy room and on a surface that will facilitate drainage of blood and fluids. The area must be clean and easily sanitized. Stainless steel necropsy tables are optimal, and some are designed with downdraft ventilation to minimize the risk of personnel exposure to hazardous agents and noxious odors. If a designated necropsy room is not feasible, an area that is isolated from other animals, personnel areas, and feed and bedding storage may be used, provided that the area can be cleaned and sanitized following each use.

In view of the tendency for ruminant carcasses to decompose, sheep and goats should be examined immediately after death. If this is not possible, the carcass should be stored in a refrigerated unit (either a large refrigerator or cold-room which is not used for storage of food for animals or personnel).

Note: Carcasses should not be frozen, as this can interfere significantly with obtaining useful data and samples from the animal.

Formaldehyde, which is commonly used in diluted form as a tissue fixative, can cause allergic reactions and irritation of surfaces lined by mucous membranes.[175] In addition, formaldehyde is considered to be a human carcinogen.[176] For these reasons, every effort should be made to limit exposure of personnel to formaldehyde, including provision of adequate ventilation of the necropsy and tissue processing areas.

Note: 37% Formaldehyde is diluted 1:9 and buffered to make 10% neutral buffered formalin.

In view of the size and weight of some small ruminant carcasses, personnel performing necropsy procedures on small ruminants should be physically fit and thoroughly familiar with published guidelines on safe lifting practices. Whenever possible, mechanical aids (such as electric hoists, if available) should be used to move carcasses to and from the necropsy table, and a gurney or dumpster used to transport the carcass from the necropsy area to the disposal site.

➤ *Necropsy Technique*

A detailed description of necropsy methods for small ruminants can be found elsewhere,[177] but general outlines are presented below:

1. The animal is first weighed and then placed on the necropsy table. The external surfaces are examined for evidence of skin discoloration, wounds, hair loss, masses, nasal or ocular discharges, and fecal or urinary staining. In addition, the oral cavity is examined, with particular attention paid to the teeth.

2. With the animal in dorsal recumbency, the skin is incised in both the axillary and inguinal regions. The forelimbs and hindlimbs are then spread out so that they lie parallel to the surface of the necropsy table, thereby supporting the carcass and maintaining it in dorsal recumbency.

3. Using a scalpel or sharp knife, the skin is incised along the ventral midline, beginning at the lower jaw and extending caudally to the pubis. Cutting from the subcutaneous side up will cause less dulling of the cutting utensil.

4. The skin is retracted gently and the subcutaneous tissues and underlying musculature exposed.

5. The abdominal wall is incised along the ventral midline, through the "white line," using the dissecting scissors.

6. The abdominal organs and the peritoneal surface are examined for abnormal discoloration, size, evidence of masses, traumatic damage, or other abnormal appearances. Depending on the time between death and necropsy, the tissues may appear abnormal because of post-mortem autolysis, a natural process involving tissue destruction by the animal's own enzymes and bacteria, and unrelated to any disease process.

7. The thoracic cavity is exposed by cutting the diaphragm and then separating the ribs from the sternum at the costochondral junction; in lambs and kids, the ribs can be cut with heavy scissors, since they are largely cartilaginous, but in mature animals bone forceps are required.

8. The lungs, heart and pleural surfaces are examined for abnormalities. The organs are then removed for more detailed examination; this can be achieved by cutting the trachea pulling the organs caudally towards the diaphragm and severing any connections between the organs and the dorsal body wall.

9. Abnormal fluids should be aspirated with a sterile syringe and needle. Body fluids should be collected into sterile glass or plastic vials and submitted for cytology, serology, virology, and/or bacterial culture, as indicated.

10. Masses or areas of abnormal tissue architecture should be measured and recorded. Samples of abnormal tissue should be removed, cut into small pieces (approximately 7-mm thickness) and fixed in 10% neutral buffered formalin. Once fixed, these tissues can be processed for histopathological examination.

notes

resources

Numerous resources are available for those seeking additional information on sheep and goats as laboratory animals. This chapter includes details of professional organizations that can be contacted for information and guidance, publications and electronic media relevant to small ruminants, and the names and contact details of vendors of animals, feed and equipment. The information is not intended to be exhaustive, but hopefully provides a starting point for interested readers.

organizations

A number of professional organizations exist which can serve as initial contacts for obtaining information regarding specific professional issues related to the care and use of laboratory sheep and goats. Membership in these organizations is recommended, since it allows the laboratory animal science professional to stay current on regulatory issues, improved procedures for the use of animals, management issues, and animal health information. The following contact details provided are correct at the time of publication:

American Association for Laboratory Animal Science (AALAS). 9190 Crestwyn Hills Drive, Memphis, TN 38125 (Tel: 901-754-8620; Fax: 901-753-0046; Internet: **www.aalas.org**). AALAS serves a diverse professional group, ranging from principal investigators to animal technicians to veterinarians. The journals, *Contemporary Topics in Laboratory Animal Science and Laboratory Animal Science*, are published by AALAS and serve to communicate relevant information. AALAS sponsors a program for certification of laboratory animal science professionals at three levels: assistant laboratory animal technician (ALAT), laboratory animal technician (LAT), and laboratory animal technologist (LATG).

The AALAS-affiliated **Institute for Laboratory Animal Management (ILAM)** provides state-of-the-art training in laboratory animal facility management. In addition, the association sponsors an annual meeting and maintains a continuing education registry for members certified at one of the specialty levels (ALAT, LAT, LATG).

The **American Association of Small Ruminant Practitioners (AASRP)**, Dee Ann Walker, 530 Church Street, Suite 700, Nashville, TN 37219. (Tel: 615-254-3687; Fax: 615-254-7047). AASRP was formed in 1968 to further education and scientific programs for veterinarians working with small ruminants. A quarterly newsletter, *Wool and Wattles*, continuing education programs at national meetings, and student externship opportunities and awards are some of the activities and membership benefits. Veterinarians, non-veterinary associates and veterinary students can be members.

The **Laboratory Animal Management Association (LAMA)**, P.O. Box 877, Killingworth, CT 06419 (Tel: 301-295-1568; Fax: 301-295-0947; Internet: **www.animalvillage.com/lama**). The membership of LAMA is comprised of professional managers, supervisors, and administrators of laboratory animal care and use programs throughout the world. LAMA's objectives are: to promote the dissemination of ideas, experience and knowledge; to encourage continued education; to act as a spokesperson on laboratory management issues; and to actively assist in the training of laboratory animal facility managers. Their quarterly publication is the *LAMA Review*.

The **American Society of Laboratory Animal Practitioners (ASLAP)**, Tel: 713-792-5127; Fax: 713-792-5796; Internet: **www.aslap.org**. ASLAP is an association of veterinarians engaged in some aspect of laboratory animal medicine. The society publishes a quarterly newsletter, sponsors a biennial continuing education conference in association with the annual AALAS meeting, and presents programs at the annual AVMA (American Veterinary Medical Association) and AALAS (American Association for Laboratory Animal Science) meetings. Membership is open to any veterinarian or veterinary student interested in the practice of laboratory animal medicine. ASLAP represents laboratory animal veterinarians through the AVMA House of Delegates.

The **American College of Laboratory Animal Medicine (ACLAM)** Dr. Mel Balk, 96 Chester Street, Chester, NH 03036 (Tel: 603-887-2467; Fax: 603-887-0096; Internet: **www.aclam.org**). ACLAM is an association of laboratory animal veterinarians founded to encourage education, training, and research in laboratory animal medicine. Laboratory Animal Medicine is recognized as a specialty of veterinary medicine, and ACLAM is the organization that certifies qualified veterinarians as Diplomates by means of examination, experience and publications in laboratory animal medicine. ACLAM hosts an annual forum, and presents programs at the annual AVMA and AALAS meetings.

The **International Council for Laboratory Animal Science (ICLAS)** Dr. Osmo Haninnen, Department of Physiology, University of Kuopio, P.O. Box 1627, SF-70211, Kuopio, Finland. ICLAS was organized to promote and coordinate the development of laboratory animal science throughout the world. ICLAS sponsors international meetings once every four years, with regional meetings held more frequently. The organization is composed of national, scientific, and union members.

The **Institute of Laboratory Animal Research (ILAR)** 2101 Constitution Avenue, NW, NAS Room 347, Washington, DC,

20418. (Tel: 202-334-2590; Fax: 202-334-1687; Internet: **www2.nas.edu/ilarhome**). The Institute of Laboratory Animal Research functions under the auspices of the National Research Council to develop and make available scientific and technical information on laboratory animals and other biological resources. Recent publications available from ILAR include, *The Guide for the Care and Use of Laboratory Animals*, *The Dog: Laboratory Animal Management*, and *Occupational Health and Safety in the Care and Use of Research Animals.*

Association for the Assessment and Accreditation of Laboratory Animal Care–International (AAALAC), 11300 Rockville Pike, Suite 1211, Rockville, MD, 20852-3035. (Tel: 301-231-5353; Fax: 301-231-8282; Internet: **www.aaalac.org).** AAALAC is a nonprofit organization which promotes the humane care and use of laboratory animals through a voluntary accreditation program. AAALAC accreditation is widely accepted as strong evidence of a quality research animal care and use program.

publications

Publications that can be referred to for additional information are listed below. This list is not all-inclusive, and mention of a reference does not indicate endorsement, nor does exclusion of a reference necessarily indicate unsuitability.

Books

1. **The Sheep as an Experimental Animal**, by J.F. Hecker. 1983. ISBN 0-12-336050-1. 216 pages. Academic Press, Inc., 525 B Street, Suite 1900, San Diego, CA 92101.

2. **The Guide for the Care and Use of Laboratory Animals**, NRC. 1996. ISBN 0-309-05377-3. 140 pages. National Academy Press, 2101 Constitution Avenue NW, Washington, DC 20418.

3. **Occupational Health and Safety in the Care and Use of Research Animals**, NRC. 1997. ISBN 0-309-05299-8.

154 pages. National Academy Press, 2101 Constitution Avenue NW, Washington, DC 20418.

4. *Formulary for Laboratory Animals*, by C.T. Hawk and S.L. Leary. 1995. ISBN 0-8138-2422-2. 101 pages. Iowa State University Press, Ames, IA 50014.

5. *Veterinary Medicine*, by D.C. Blood and J.A. 1994. Henderson. ISBN 0-702-01592-X. 1787 pages. 8th edition, W.B. Saunders Company, The Curtis Center, Independence Square West, Philadelphia, PA 19106.

6. *Domestic Animal Behavior for Veterinarians and Animal Scientists,* by K.A. Houpt. 1998. ISBN 0-8138-1060-4. 356 pages. Iowa State University Press, Ames, IA 50014.

7. *Anesthesia and Analgesia in Laboratory Animals,* edited by D.F. Kohn, S.K. Wixson, W.J. White, and G.J. Benson. 1997. ISBN 0-12-417570-8. 426 pages. Academic Press, 15 East 26th Street, 15th Floor, New York, NY 10010.

8. *Atlas of Topographic Anatomy of the Domestic Animals,* by P. Popesko. 1978. ISBN: 0-72-167275-2. 207 pages. W.B. Saunders Company, The Curtis Center, Independence Square West, Philadelphia, PA 19106.

9. *The UFAW Handbook on the Care and Management of Laboratory Animals,* edited by T. Poole. 6th edition. 1987. ISBN: 0-582-40911-X. 933 pages. Longman Scientific & Technical. Burnt Mill, Harlow, UK.

10. *Jensen and Swift's Diseases of Sheep,* by C.V. Kimberling. 1988. ISBN 0-8121-1099-4. 3th edition. 394 pages. Lea & Febiger, 600 Washington Square, Philadelphia, PA 19106.

Periodicals

Periodicals are a valuable source of information on the latest techniques or trends. Some periodicals that often have articles on sheep and goats as laboratory animals include:

1. ***Laboratory Animal Science.*** Refer to AALAS under **Organizations** above for contact and subscription information.

2. ***Contemporary Topics in Laboratory Animal Science.*** Refer to AALAS under **Organizations** above for contact and subscription information.

3. ***Laboratory Animals.*** Published monthly by Lab Animals Ltd. (Tel: +44 171 290 2927; Internet: http://www.lal.org.uk).

4. ***Lab Animal.*** Published by the Nature Company. Internet: http://www.labanimal.com).

5. ***AASRP Newsletter.*** Refer to AASRP under **Organizations** above for contact and subscription information.

6. ***ILAR Journal.*** Refer to ILAR under **Organizations** above for contact and subscription information.

7. ***Small Ruminant Research.*** Official journal of the International Goat Association. The journal is published by Elsevier Science. Internet: http://www.elsevier.com).

electronic resources

Information resources available through the computer continue to increase in quality and quantity. The following sites can be referred to for electronic resources or information related to sheep and goats as laboratory animals.

1. Breeds of Sheep
   http://www.ansi.okstate.edu/breeds/sheep/

2. Goat Breeds
   http://www.ansi.okstate.edu/breeds/goats/

3. American Association for Laboratory Animal Science
   http://www.aalas.org

4. National Pygmy Goat Association
   http://www.tiac.net/users/npga/

5. Anesthetic/Analgesics in Small Ruminants
    http://www.med.stanford.edu/school/Comp
    Med/Resource/anesthetics-rum.html

6. Sheep/Goat Resources
    http://www.aps.uoguelph.ca/~lohuism/sheepgoat.html

7. Sheep Listserver
    http://www.tile.net/tile/listserv/sheepl.html

8. American Sheep Industry Association
    http://sheepusa.org/

9. USDA Scrapie Factsheet
    http://www.aphis.usda.gov/oa/scrapie.html

10. Goat Resources
    http://www.ansi.okstate.edu/library/goats.html

11. Sheep Resources
    http://www.ansi.okstate.edu/library/sheep.html

12. Sheep Brain Dissection Guide
    http://academic.uofs.edu/department/psych/sheep

13. Sheep Brain Atlas
    http://web.mit.edu/afs/
    athena.mit.edu/org/b/bcs/www/sheepatlas/sheep.htm

14. Extension Publications on Sheep
    http://www.ianr.unl.edu/pubs/Sheep

15. Lab Animal Buyer's Guide
    http://guide.labanimal.com

sources of sheep and goats as laboratory animals

In contrast to some other laboratory animals, such as rodents and rabbits, sheep and goats are not as readily available from commercial vendors. Although this may make obtaining quality research animals more challenging, time spent initially identifying and verifying a high-quality vendor is always worth the time. Vendors should be asked to supply information on the heath status of their animals prior to purchase; as a minimum,

this information should include details of the routine vaccination and anthelmintic programs. Specific serological testing (e.g.,
for CAEV, Q fever, or OPP) is advisable when considering animals
for chronic studies. Attention should also be paid to the transportation of sheep and goats, as the stress of shipping may
activate latent diseases such as pneumonia and contagious
ecthyma.

Lab Animal publishes an annual buyer's guide in December
of each year, and this should be referred to for potential sources
of animals. Additionally, the American Association for Laboratory
Animal Science (AALAS) publishes a print and electronic resource
directory. If you are located in a rural area, or there are farms
or ranches or a university with an agricultural component
nearby, it may be possible to utilize local sources of sheep and
goats. This has the advantage of decreasing the cost and stress
of shipping, and the cost of the animals may be competitive with
or lower than other sources. Regardless of the source, attention
needs to be paid to the quality and reliability of the animal.
Subclinical pneumonia is *very* common in sheep flocks, and may
invalidate studies, or complicate anesthesia and surgery.

It is impractical to list all vendors here, but the following are
examples of commerical sources of sheep and/or goats.

1. Animal Biotech Industries, P.O. Box 519, Danboro, PA
   18916 (Tel: 1-215-766-7413). Sheep and goats.

2. Archer Farms, P.O. Box 170, Belcamp, MD 21017 (Tel:
   410-879-4110). Sheep and goats.

3. Barton's West End Farms, Inc., 161 Janes Chapel Road,
   Oxford, NJ 07863 (Tel: 908-637-4427). Sheep and goats.

4. Buckshire Corporation, 2025 Ridge Rd, Perkasie, PA
   18944 (Tel: 215-257-0116). Sheep and goats.

5. Middlefork Kennels, Box 64, Salisbury, MO 65281 (Tel:
   816-388-5860). Sheep and goats.

6. OVIS, 47825 279th Street, Canton, SD 57013 (Tel: 605-
   987-4402). Sheep.

7. Thomas D. Morris, Inc., 4001 Millender Mill Road, Reis-
   terstown, MD 21136 (Tel: 410-356-6780).

8. West Jersey Bio-Services Inc., P.O. Box 6, Wenonah, NJ 08090 (Tel: 609-468-1776). Sheep and goats.

feed

Rations for sheep and goats should include a source of roughage as well as an energy source. Hay is the most common source of roughage for small ruminants. The quality of hay supplies can be tremendously variable, and it is prudent to use a reputable local source whenever possible. Hay that is stored in barns may become contaminated by droppings from wild rodents, birds and cats, and may act as a source of disease if used in a laboratory animal facility. It is also advisable to analyze the hay for total digestible nutrients, total protein, and trace mineral concentrations.

Feed for energy (concentrates) can be obtained from suppliers of laboratory animal diets, as listed below. Additionally, these are usually available from local feed mills, although the latter may be of more variable quality than diets from vendors that routinely manufacture and distribute laboratory animal diets. Analysis of these feeds is recommended initially, and at least annually thereafter. It is important to remember that sheep should not be fed cattle rations, since the latter contain levels of copper that may be toxic to sheep.

1. Harlan Teklad, Inc., P.O. Box 44220, Madison, WI 53744 (Tel: 608-277-2066).

2. PMI/Purina Mills, Inc., 505 North 4th St., P.O. Box 548, Richmond, IN 47375 (Tel: 1-800-227-8941).

3. Dyets, Inc., 2508 Easton Avenue, Bethelem, PA 18017 (Tel: 1-800-275-3938).

4. Nebeker Ranch, 1639 12th Street, Santa Monica, CA 90404 (Tel: 310-450-1334).

5. ICN Pharmaceuticals, Biomedical Research Products, 3300 Hyland Avenue, Costa Mesa, CA 92626 (Tel: 1-800-854-0530).

equipment

Sanitation Supplies

Several sources of disinfectants and other sanitation supplies are listed below:

1. Pharmacal Research Labs, Inc., P.O. Box 369, Naugatuck, CT 06770 (Tel: 1-800-243-5350).

2. BioSentry, Inc., 1481 Rock Mountain Blvd., Stone Mountain, GA 30083 (Tel: 1-800-788-4246).

3. Calgon Vestal Contamination Control, P.O. Box 147, St. Louis, MO 63166 (Tel: 1-800-582-6514).

4. Rochester Midland, Inc., 333 Hollenbeck St., P.O. Box 1515, Rochester, NY 14603 (Tel: 1-800-836-1627).

Cages and Research and Veterinary Supplies

A number of sources of pharmaceuticals, hypodermic needles, syringes, surgical equipment, bandages, and other related items are provided below. Pharmaceuticals should generally be ordered and used only under the direction of a licensed veterinarian. Cages and pens should meet the size requirements as specified by relevant regulatory agencies.

TABLE 18. POSSIBLE SOURCES OF EQUIPMENT AND SUPPLIES

Item	Source
Cages and supplies	1, 3, 5, 7, 9, 10, 12
Veterinary and surgical supplies	4, 7, 8, 13, 15
Gas anesthesia equipment	3, 7, 8, 13
Handling equipment	11, 13
Syringes and needles	4, 6, 8, 15
Vascular access equipment	3, 7
Osmotic pumps	2
Necropsy equipment	6, 7, 8, 9, 14

Contact Information for Cages, Research, and Veterinary Supplies

1. Alternative Design Manufacturing & Supply, Inc., 16396 Old Highway 68, Siloam Springs, AR 72761 (Tel: 1-800-320-2459).

2. Alza Corporation, 950 Page Mill Rd., P.O. Box 10950, Palo Alto, CA 94303 (Tel: 650-962-2251).

3. Braintree Scientific, Inc., P.O. Box 361, Braintree, MA 02184 (Tel: 617-843-2202).

4. Butler Co., Inc., 5000 Bradenton Ave., Dublin, OH 43017 (Tel: 1-800-225-7911).

5. Fenco Cage Products, 118 Dorchester Ave, Boston, MA 02125 (Tel: 617-265-9000).

6. Fisher Scientific, Inc., 711 Forbes Ave., Pittsburgh, PA 15219 (Tel: 1-800-766-7000).

7. Harvard Apparatus, Inc., 22 Pleasant St., South Natick, MA 01760 (Tel: 508-655-7000).

8. J.A. Webster, Inc., 86 Leominster Road, Sterling, MA 01564 (Tel: 1-800-225-7911).

9. Lab Products, Inc., 255 West Spring Valley Ave., P.O. Box 808, Maywood, NJ 07607 (Tel: 1-800-526-0469).

10. Lenderking Caging Products, 8370 Jumpers Hole Rd., Millersville, MD 21108 (Tel: 410-544-8795).

11. LGL Animal Care Products Inc., 1520 Cavitt St., Bryan, TX 77801 (Tel: 409-775-1776).

12. Modular Systems, Inc., 91 Spindle Point Rd., Meredith, NH 03253 (Tel: 603-279-6278).

13. Pipestone Veterinary Supply, Pipestone, MN 56164 (Tel: 507-825-4211).

14. Shandon, Inc., 171 Industry Drive, Pittsburgh, PA 15275 (Tel: 1-800-245-6212).

15. Viking Medical, P.O. Box 2142, Medford Lakes, NJ 08055 (Tel: 609-953-0138).

notes

bibliography

1. United States Department of Agriculture, *Animal Welfare Enforcement: Fiscal Year 1996*, Washington, D.C., 1997.

2. French, M. H., *Observations on the Goat*, Food and Agriculture Organization (FAO), Rome, Italy, 1970.

3. Lynch, J. J., Hinch, G. N., and Adams, D. B., *The Behaviour of Sheep*, C.A.B. International, Oxford, UK, 1992.

4. FAO, *Animal Production Yearbook*, Food and Agricultural Organization, Rome, Italy, 1992.

5. Knights, M. and Garcia, G. W., The status and characteristics of the goat (*Capra hircus*) and its potential role as a significant milk producer in the tropics: a review, *Small Ruminant Research* 26, 203–215, 1997.

6. Hafez, E. S. E., Cairns, R. B., Hulet, C. V. and Scott, J. P., The behaviour of sheep and goats, in *The Behaviour of Domestic Animals*, Hafez, E.S.E., Ed., Bailliere Tindall, London, UK, 1969, 296–308.

7. Houpt, K. A., *Domestic Animal Behavior for Veterinarians and Animal Scientists*, Iowa State University Press, Ames, IA, 1998.

8. Smith, M.C. and Sherman, D., *Goat Medicine*, Lea & Febiger, Philadelphia, PA, 1994.

9. Spence, J. and Aitchison, G., Clinical aspects of dental disease in sheep, in *Sheep and Goat Practice*, Boden, E., Ed., Bailliere Tindall Ltd, London; UK, 1991, 133–155.

10. Williams, C. S. F., Goat, in *Practical Guide to Laboratory Animals*, The C.V. Mosby Company, St. Louis, MO, 1976, 93–108.

11. Thirkell, E. J., Lanyon, M. and Strickland, N. C., Disbudding and descenting goats: anatomical considerations, *Goat Veterinary Society Journal* 11, 66–68, 1990.

12. Williams, C. S. F., Sheep, in *Practical Guide to Laboratory Animals*, The C.V. Mosby Company, St. Louis, MO, 1976, 109–122.

13. Wolfensohn, S. and Lloyd, M., *Handbook of Laboratory Animal Management and Welfare*, Oxford University Press, Oxford, UK, 1994.

14. Hecker, J. F., *The Sheep as an Experimental Animal*, Academic Press, New York, NY, 1983.

15. Mohammed, H. H. and Owen, E., Comparison of maintenance energy requirements of sheep and goats, *Animal Production* 30, 479, 1980.

16. National Research Council, *Nutrient Requirements of Sheep*, 6th edition, National Academy Press, Washington, D.C., 1985.

17. *The Goat Extension Handbook*, National Dairy Database, University of Wisconsin. Available via the Internet at http://www.inform.umd.edu/EdRes/Topic/AgrEnv/ndd/goat/

18. Brooks, D. L., Tillman, P. C. and Niemi, S. M., Ungulates as laboratory animals, in *Laboratory Animal Medicine*, Fox, J. G., Cohen, B. J., and Loew, F. M., Eds., Academic Press, Inc., Orlando, FL, 274–295.

19. Reece, W. O., The Kidneys, in *Duke's Physiology of Domestic Animals*, 11th edition, Swenson, M. J. and Reece, W. O., Eds., Cornell University Press, Ithaca, NY, 1993, 573–603.

20. Davies, D. M. and Sims, B. J., Welsh and Marches Goat Society survey to determine normal blood biochemistry and haematology in domestic goats, *Goat Veterinary Society Journal* 6, 38–42, 1985.

21. Clarkson, M. J. and Faull, W. B., *A Handbook for the Sheep Clinician*, 4th edition, Liverpool University Press, Liverpool, UK, 1990.

22. Carlson, G. P., Fluid, electrolyte, and acid-base balance, in *Clinical Biochemistry of Domestic Animals*, Kaneko, J. J., Harvey, J. W., and Bruss, M. L., Eds., 5th edition, Academic Press, San Diego, CA, 1997, 496.

23. Plumb, D. C., *Veterinary Drug Handbook Pocket Edition*, Iowa State University Press, Ames, IA, 1995.

24. Kaneko, J. J., Harvey, J. W. and Bruss, M. L., *Clinical Biochemistry of Domestic Animals*, 5th edition, Academic Press, San Diego, CA, 1997.

25. Altman, P. L. and Dittmer, D. S., *Biological Data Book*, Volume III, 2nd edition, Federation of American Societies for Experimental Biology, Bethesda, MD, 1974, 1978–1979.

26. Brewer, B. D., Neurologic disease of sheep and goats, *Veterinary Clinics of North America, Large Animal Practice* 5, 677–700, 1983.

27. Scott, P. R., Sargison, N. D., Penny, C. D., Pirie, R. S. and Kelly, J. M., Cerebrospinal fluid and plasma glucose concentrations of ovine pregnancy toxaemia cases, in appetent ewes and normal ewes during late gestation, *British Veterinary Journal* 151, 39–44, 1995.

28. Lal, S. B., Swarup, D., Dwivedi, S. K. and Sharma, M. C., Biochemical alterations in serum and cerebrospinal fluid in experimental acidosis in goats, *Research in Veterinary Science* 50, 208–210, 1991.

29. Hales, J. R. S. and Webster, M. E. D., Respiratory function during thermal tachypnea in sheep, *Journal of Physiology (London)* 190, 241–260, 1967.

30. Bakima, M., Gustin, P., Lekeux, P. and Lomba, F., Mechanics of breathing in goats, *Research in Veterinary Science* 45, 332–336, 1988.

31. Lahiri, S., Blood oxygen affinity and alveolar ventilation in relation to body weight in mammals, *American Journal of Physiology* 229, 529–536, 1975.

32. Heard, D. J., Comparative cardiopulmonary effects of intramuscularly administered etorphine and carfentanil in goats, *American Journal of Veterinary Research* 57, 87–96, 1996.

33. Halmagi, D. F. J. and Gillett, D. J., Cardiorespiratory consequences of corrected gradual severe blood loss in unanaesthetized sheep, *Journal of Applied Physiology* 21, 589–596, 1966.

34. Carroll, G. L., Hooper, R. N., Slater, M. R., Hartsfield, S. M. and Matthews, S. M., Detomidine-butorphanol-propofol for carotid artery translocation and castration or ovariectomy in goats, *Veterinary Surgery* 27, 75–82, 1998.

35. Lagutchik, M. S., Januszkiewicz, A. J., Dodd, K. T. and Martin, D. G., Cardiopulmonary effects of a tiletamine-zolazepam combination in sheep, *American Journal of Veterinary Research* 52, 1441–7, 1991.

36. Jain, N. C., Comparative hematology of common domestic animals, in *Essentials of Veterinary Hematology*, Jain, N. C., Ed., Lea & Febiger, Philadelphia, 1993, 19–53.

37. Torrington, K. G., McNeil, J. S., Phillips, Y. Y. and Ripple, G. R., Blood volume determinations in sheep before and after splenectomy, *Laboratory Animal Science* 39, 598–602, 1989.

38. National Research Council, *Guide for the Care and Use of Laborary Animals*, National Academy Press, Washington, D.C., 1996.

39. National Research Council, *Guide for the Care and Use of Agricultural Animals in Agricultural Research and Teaching*, National Academy Press, Washington, D.C., 1996.

40. Animal Welfare Act, United States P.L. 89-544, 1966; P.L. 91-579, 1970; P.L. 94-279, 1976; and P.L. 99-198, 1985 (The Food Security Act).

41. Henderson, D. C., Manipulation of the breeding season in goats: a review, *Goat Veterinary Society Journal* 8, 7–16, 1987.

42. Kimberling, C. V., *Jensen and Swift's Diseases of Sheep*, 3rd edition, Lea & Febiger, Philadelphia, PA, 1988.

43. Qi, K. and Huston, J. E., A review of thiamin requirement and deficiency of sheep and goats, *Sheep and Goat Research Journal* 11, 25–30, 1995.

44. Martin, B. J., Dysko, R. C., Chrisp, C. E. and Ringler, D. H., Copper poisoning in sheep, *Laboratory Animal Science* 38, 734–736, 1988.

45. National Research Council, *Nutrient Requirements of Goats*, National Academy Press, Washington, D.C., 1981.

46. Code of Federal Regulations, Title 9, Parts 1, 2, and 3 (Docket 89-130) *Federal Register* 54 (168), 1989.

47. Code of Federal Regulations, Title 9, Part 3 (Docket 90-218), *Federal Register* 56 (32), 1991.

48. Arthur, G. H., Noakes, D. E. and Pearson, H., *Veterinary Reproduction and Obstetrics*, W. B. Saunders, London, UK, 1996.

49. Henderson, D. C., Control of the breeding season in sheep and goats, in *Sheep and Goat Practice*, Boden, E., Ed., Bailliere Tindall Ltd, London, UK, 1991, 133–155.

50. Allison, A. J. and Kelly, R. W., Synchronisation of oestrus and fertility in sheep treated with progestogen-impregnated implants, and prostaglandins with or without intravaginal sponges and subcutaneous pregnant mare's serum, *New Zealand Journal of Agricultural Research* 21, 389–393, 1978.

51. Goel, A. K. and Agrawal, K. P., A review of pregnancy diagnosis techniques in sheep and goats, *Small Ruminant Research* 9, 255–264, 1992.

52. Dawson, L. J., Sahlu, T., Hart, S. P., Detweiler, G., Gipson, T. A., Teh, T. H., Henry, G. A. and Bahr, R. J., Determination of fetal numbers in Alpine does by real-time ultrasonography, *Small Ruminant Research* 14, 225–231, 1994.

53. Edqvist, L.-E. and Forsberg, M., Clinical reproductive endocrinology, in *Clinical Biochemistry of Domestic Animals*, Kaneko, J. J., Harvey, J. W. and Bruss, M. L., Eds., 5th edition, Academic Press, San Diego, CA, 1997, 606.

54. Peters, A. R. and Dent, C. N., Induction of parturition in sheep using dexamethasone, *Veterinary Record* 131, 128–129, 1992.

55. Rook, J. S., Scholman, G., Wing-Proctor, S. and Shea, M., Diagnosis and control of neonatal losses in sheep, *Veterinary Clinics of North America (Food Animal Practice)* 6, 531–562, 1990.

56. Kent, J. E., Molony, V. and Robertson, I. S., Comparison of the Burdizzo and rubber ring methods for castrating and tail docking lambs, *Veterinary Record* 136, 192–196, 1995.

57. Mackenzie, D., Goat Husbandry, in *Goat Husbandry*, Faber & Faber, London, UK, 1993, 109–112.

58. Health Research Extension Act, United States P.L. 99–158, 1985.

59. Office for Protection of Research Risks, *Public Health Service Policy on Humane Care and Use of Laboratory Animals*, National Institutes of Health, Bethesda, MD, 1986.

60. National Research Council, *Occupational Health and Safety in the Care and Use of Research Animals*, National Academy of Sciences, Washington, D.C., 1997.

61. Guo, H. R., Tanaka, S., Cameron, L. L., Seligman, P. J., Behrens, V. J., Ger, J., Wild, D. K. and Putz-Anderson, V., Back pain among workers in the United States: national estimates and workers at high risk, *American Journal of Industrial Medicine* 28, 591–602, 1995.

62. National Institute for Occupational Safety and Health, *Work Practices Guide for Manual Lifting*, United States Government Printing Office, Washington, D.C., 1991.

63. Slavin, R. G., Contact dermatitis, in *Allergic Diseases: Diagnosis and Management*, Patterson, R., Grammer, L. C., Greenberger, P. A. and Zeiss, C. R., Eds., J. B. Lippencott, Philadelphia, PA, 1993, 553–558.

64. Fuortes, L. J., Weih, L., Jones, M. L., Burmeister, L. F., Thorne, P. S., Pollen, S. and Merchant, J. A., Epidemiologic assessment of laboratory animal allergy among university employees, *American Journal of Industrial Medicine* 29, 67–74, 1996.

65. Blainey, A. D., Topping, M. D., Ollier, S., and Davies, R. J., Respiratory symptoms in arable farmworkers: role of storage mites, *Thorax* 43, 697–702, 1988.

66. Center for Disease Control-National Institutes of Health, *Biosafety in Microbiological and Biomedical Laboratories*, United States Government Printing Office, Washington, D.C., 1993.

67. Huerter, C. J., Alvarez, L., and Stinson, R., Orf: case report and literature review, *Cleveland Clinic's Journal of Medicine* 58, 531–534, 1991.

68. Glaser, C. A., Angulo, F. J., and Rooney, J. A., Animal-associated opportunistic infections among persons infected with the human immunodeficiency virus, *Clinical Infectious Diseases* 18, 14–24, 1994.

69. Kampelmacher, E. H. and van Noorle Jansen, L. M., Listeriosis in humans and animals in the Netherlands (1958–1977), *Zentralblatt fur Bakteriologie, Reihe A*, 246, 211–27, 1980.

70. Blewett, D., The epidemiology of ovine toxoplasmosis, *Veterinary Annual* 25, 120–124, 1985.

71. Dubey, J. P., Toxoplasmosis, *Journal of the American Veterinary Medical Association*, 205, 1593–8, 1994.

72. Amin, J. D. and Wilsmore, A. J., Studies on the early phase of the pathogenesis of ovine enzootic abortion in the nonpregnant ewe, *British Veterinary Journal* 151, 141–155, 1995.

73. Helm, C. W., Smart, G. E., Cumming, A. D., Lambie, A. T., Gray, J. A., MacAulay, A., and Smith, I. W., Sheep-acquired severe *Chlamydia psittaci* infection in pregnancy, *International Journal of Gynaecology and Obstetrics* 28, 369–72, 1989.

74. Behymer, D. and Riemann, H. P., *Coxiella burnetii* infection (Q fever), *Journal of the American Veternary Medical Association* 194, 764–7, 1989.

75. Bernard, K. W., Parham, G. L., Wunkler, W. G., and Helmick, C. G., Q fever control measures: recommendations for research facilities using sheep, *Infection Control and Hospital Epidemiology* 3, 6, 1982.

76. Singh, S. B. and Lang, C. M., Q fever serological surveillance program for sheep and goats at a research animal facility, *American Journal of Veterinary Research* 46, 321–325, 1985.

77. Russel, A., Body condition scoring of sheep, in *Sheep and Goat Practice*, Boden, E., Ed., Bailliere Tindall Ltd., London, UK, 1991, 3–10.

78. Morgan, K., Footrot, in *Sheep and Goat Practice*, Boden, E., Ed., Bailliere Tindall Ltd., London, UK, 1991, 167–182.

79. Radostits, O. M., Blood, D. C., and Gay, C. C., *Veterinary Medicine: A Textbook of the Diseases of Cattle, Sheep, Pigs, Goats and Horses*, 8th edition, Balliere Tindall Ltd., London, UK, 1994.

80. Bennett, D. G., Parasites of the respiratory system, in *Current Veterinary Therapy 2: Food Animal Practice*, Howard, J. L., Ed., W. B., Saunders, Co., Philadelphia, 1986, 684–687.

81. Brodie, S. J., de la Concha-Bermejillo, A., Snowder, G. D., and DeMartini, J. C., Current concepts in the epizootiology, diagnosis, and economic importance of ovine progressive pneumonia in North America: a review, *Small Ruminant Research* 27, 1–17, 1998.

82. Cutlip, R. C., Lehmkuhl, H. D. and Brogden, K. A., Seroprevalence of ovine progressive pneumonia virus in various domestic and wild animal species, and species susceptibility to the virus, *American Journal of Veterinary Research* 52, 189–191, 1991.

83. Urquhart, G. M., Armour, J., Duncan, J. L., Dunn, A. M., and Jennings, F. W., *Veterinary Parasitology*, Longman Scientific & Technical, Harlow, UK., 1992.

84. Coles, G., Parasite control in sheep, *In Practice* 16, 309–318, 1994.

85. Penzhorn, B. L. and Swan, G. E., Coccidiosis, in *Current Veterinary Therapy 3: Food Animal Practice*, Howard, J. L., Ed., W. B. Saunders, Co., Philadelphia, 1993, 599–604.

86. Parajuli, B. and Goddard, P. J., A comparison of the efficacy of footbaths containing formalin or zinc sulphate in treating ovine foot-rot under field conditions, *British Veterinary Journal* 145, 467–472, 1989.

87. Wyman, M., Eye disease of sheep and goats, *Veterinary Clinics of North America (Food Animal Practice)* 5, 657–675, 1983.

88. Dagnall, G. J. R., The role of *Branhamella ovis*, *Mycoplasma conjunctivae*, and *Chlamydia psittaci* in conjunctivitis of sheep, *British Veterinary Journal* 150, 65–71, 1994.

89. Sakul, H., Snowder, G. D., and Hemenway, K. J., Evaluation of techniques for correction of entropion in lambs, *Small Ruminant Research* 20, 187–191, 1996.

90. Sakul, H. and Kellom, T. R., Heritability of entropion in several U.S. sheep breeds, *Small Ruminant Research* 23, 187–190, 1997.

91. Harwood, D., Common goat ailments, *Goat Veterinary Society Journal* 15, 117–126, 1994.

92. Rakestraw, P. C., Fubini, S. L., Gilbert, R. O., and Ward, J. O., Tube cystostomy for treatment of obstructive urolithiasis in small ruminants, *Veterinary Surgery* 24, 498–505, 1995.

93. East, N. E., Rowe, J. D., Dahlberg, J. E., Theilen, G. H. and Pedersen, N. C., Modes of transmission of caprine arthritis-encephalitis virus infection, *Small Ruminant Research* 10, 251–262, 1993.

94. Wilkerson, M. J., Davis, W. C., and Cheevers, W. P., Peripheral blood and synovial fluid mononuclear cell phenotypes in lentivirus-induced arthritis, *Journal of Rheumatology* 22, 8–15, 1995.

95. Prusiner, S. B., Prions and neurodegenerative diseases, *New England Journal of Medicine* 317, 1571–1581, 1987.

96. Detwiler, L. A., Jenny, A. L., Rubenstein, R., and Wineland, N. E., Scrapie: a review, *Sheep & Goat Research Journal* 12, 111–131, 1996.

97. Wang, S., Maciulis, A., Holyoak, G. R., Foote, W. C., Clark, W., and Bunch, T. D., A test of EcoRI and HindIII restriction fragment length polymorphisms in assessing susceptibility for scrapie in U.S. Suffolk sheep, *Small Ruminant Research* 28, 123–130, 1998.

98. Appleyard, B. and Bailie, H., Parasitic skin diseases of sheep, in *Sheep and Goat Medicine*, Boden, E., Ed., Bailliere Tindall Ltd., London, UK, 1991, 211–226.

99. Jackson, P., Skin diseases in goats, in *Sheep and Goat Medicine*, Boden, E., Ed., Bailliere Tindall Ltd., London, UK, 1991, 248–265.

100. Mitra, M., Mahanta, S. K., Sen, S., Ghosh, C. and Hati, A. K., Transmission of *Sarcoptes scabiei* from animal to man and its control, *Journal of the Indian Medical Association* 93, 142–143, 1995.

101. Bretzlaff, K., Production medicine and health programs for goats, in *Current Veterinary Therapy 3: Food Animal Practice*, Howard, J. L., Ed., W. B. Saunders, Co., Philadelphia, 1993, 162–167.

102. Lloyd, S., Caseous lymphadenitis in sheep and goats, *In Practice* 16, 24–29, 1994.

103. Hamliri, A., Kessabi, M., Johnson, D. W., and Olson, W. G., Prevention of nutritional myopathy in sheep grazing selenium-deficient pastures, *Small Ruminant Research* 10, 13–23, 1993.

104. Jenkins, W. L., Antimicrobial therapy, in *Current Veterinary Therapy 2: Food Animal Practice*, Howard, J. L., Ed., W. B. Saunders, Co., Philadelphia, 1986, 8–23.

105. Naxcel Product Insert, Upjohn-Pharmacia, 1997.

106. Moore, L. F. and Shmidl, J. A., External parasiticides, in *Current Veterinary Therapy 3: Food Animal Practice*, Howard, J. L., Ed., W. B. Saunders Co., Philadelphia, 1993, 51–58.

107. Courtney, C. H. and Roberson, E. L., Anticestodal and antitrematodal drugs, in *Veterinary Pharmacology and Therapeutics*, Adams, H. R., Ed., 7th edition, Iowa State University Press, Ames, IA, 1995, 933–954.

108. Courtney, C. H. and Roberson, E. L., Anti-nematodal agents, in *Veterinary Pharmacology and Therapeutics*, Adams, H. R., Ed., Iowa State University Press, Ames, IA, 1995, 885–932.

109. Brander, C. G., Pugh, D. M., and Bywater, R. J., *Veterinary Applied Pharmacology and Therapeutics*, Bailliere Tindall, London, UK, 1982.

110. Hall, L. W. and Clarke, K. W., *Veterinary Anaesthesia*, Bailliere Tindall, London, UK, 1991.

111. Short, C. E., *Principles & Practice of Veterinary Anesthesia*, Williams & Wilkins, Baltimore, MD, 1987.

112. Thurmon, J. C., Tranquilli, W. J., and Benson, G. J., *Lumb and Jones' Veterinary Anesthesia*, 3rd edition, Williams and Wilkins, 1996).

113. Thurmon, J. C., Injectable anesthetic agents and techniques in ruminants and swine, *Veterinary Clinics of North America (Food Animal Practice)* 2, 567–591, 1986.

114. Eissa, H. M., Ahmed, A. S., and El-Sayed, M. A. I., Influence of general anaesthesia (thiopental sodium) on progesterone and cortisol levels in pregnant goat serum, *Archiv fur Experimentelle Veterinarmedizin* 44, 621–625, 1990.

115. Flecknell, P., Anaesthesia of common laboratory species, in *Laboratory Animal Anaesthesia. An Introduction for Research Workers and Technicians*, Flecknell, P., Ed., Academic Press, London, 1987, 107–109.

116. Nowrouzian, I., Schels, H. F., Ghodsian, I., and Karimi, H., Evaluation of the Anaesthetic properties of ketamine and a ketamine/xylazine/atropine combination in sheep, *Veterinary Record* 108, 354–356, 1981.

117. Kumar, A., Thurmon, J. C., and Hardenbrook, H. J., Clinical studies of ketamine HCl and xylazine HCl in domestic goats, *Veterinary Medicine and Small Animal Clinics* 71, 1707–1713, 1976.

118. Reid, J., Nolan, A. M., and Welsh, E., Propofol as an induction agent in the goat: a pharmacokinetic study, *Journal of Veterinary Pharmacology and Therapeutics* 16, 488–493, 1993.

119. Waterman, A. E., Use of propofol in sheep, *Veterinary Record* 122, 260, 1988.

120. Lin, H. C., Tyler, J. W., Wallace, S. S., Thurmon, J. C., and Wolfe, D. F., Telazol and xylazine anesthesia in sheep, *Cornell Veterinarian* 83, 117–124, 1993.

121. Hall, L. W., Althesin in the larger animal, *Postgraduate Medical Journal* 48 Suppl. 2, 55–58, 1972.

122. Antognini, J. F. and Eisele, P. H., Anesthetic potency and cariopulmonary effects of enflurane, halophane, and isoflurane in goats, *Laboratory Animal Science* 43, 607–610, 1993.

123. Palahniuk, R. J., Schnider, S. M., and Eger II, E. I., Pregnancy decreases the requirement for inhaled anesthetic agents, *Anesthesiology* 41, 82-83, 1974.

124. Skarda, R. R., Local and regional anesthesia in ruminants and swine, *Veterinary Clinics of North America (Food Animal Practice)* 12, 579–626, 1996.

125. Babalola, G. O. and Oke, B. O., Intravenous regional analgesia for surgery of the limbs in goats, *Veterinary Quarterly* 5, 186–189, 1983.

126. Brock, K. A. and Heard, D. J., Field anesthesia techniques in small ruminants. I. Local analgesia, *Compendium on Continuing Education for the Practicing Veterinarian* 7, S417–S424, 1985.

127. Pablo, L. S., Epidural morphine in goats after hindlimb orthopedic surgery, *Veterinary Surgery* 22, 307–310, 1993.

128. Coombs, D. W., Colburn, R. W., DeLeo, J. A., Hoopes, P. J., and Twitchell, B. B., Comparative histopathology of epidural hydrogel and silicone elastomer catheters following 30 and 180 days implant in the ewe, *Acta Anaesthesiologica Scandinavica* 38, 388–95, 1994.

129. Gray, P. R. and McDonell, W. N., Anesthesia in goats and sheep. I. Local analgesia, *Compendium of Continuing Education for the Practicing Veterinarian* 8, S33–S39, 1986.

130. Singh, J., Singh, A. P., Peshin, P. K., Sharif, D., and Patil, D. B., Evaluation of detomidine as a sedative in sheep, *Indian Journal of Animal Sciences* 64, 237–238, 1994.

131. Trim, C. M., Special anesthesia considerations in the ruminant, in *Principles & Practice of Veterinary Anesthesia*, Short, C. E., Ed., Williams & Wilkins, Baltimore, MD, 1987, 285–300.

132. Morton, D. B. and Griffiths, P. H. M., Guidelines on the recognition of pain, distress and discomfort in experimental animals and an hypothesis for assessment, *Veterinary Record* 116, 431–436, 1985.

133. Dowd, G., Gaynor, J. S., Alvis, M., Salman, M., and Turner, A. S., A comparison of transdermal fentanyl and oral phenylbutazone for postoperative analgesia in sheep. *Veterinary Surgery* 27, 168, 1997.

134. Welsh, E. M., Gettinby, G., and Nolan, A. M., Comparison of a visual analogue scale and a numerical rating scale for assessment of lameness, using sheep as a model, *American Journal of Veterinary Research* 54, 976–983, 1993.

135. Waterman, A. E., Livingston, A. and Amin, A., The antinociceptive activity and respiratory effects of fentanyl in sheep, *Journal of the Association of Veterinary Anaesthetists of Great Britain and Ireland*, 17, 21–23, 1990.

136. Kopcha, M., Kaneene, J. B., Shea, M. E., Miller, R. A., and Ahl, A. S., Use of nonsteroidal anti-inflammatory drugs in food animal practice, *Journal of the American Veterinary Medical Association* 201, 1868–1872, 1992.

137. Welsh, E. M. and Nolan, A. M., Effect of flunixin meglumine on the thresholds to mechanical stimulation in healthy and lame sheep, *Research in Veterinary Science* 58, 61–66, 1995.

138. Welsh, E. M. and Nolan, A. M., Effects of non-steroidal anti-inflammatory drugs on the hyperalgesia to noxious mechanical stimulation induced by the application of a tourniquet to a forelimb of sheep, *Research in Veterinary Science* 57, 285–291, 1994.

139. Scott, P. R. and Gessert, M. E., Management of post-partum cervical uterine or rectal prolapses in ewes using caudal epidural xylazine and lignocaine injection, *British Veterinary Journal* 153, 115–116, 1997.

140. Randolph, M. M., Post-operative care and analgesia of farm animals used in biomedical research, *Animal Welfare Information Center Newsletter* 5, 11–13, 1994.

141. Waterman, A. E., Livingston, A., and Amin, A., Analgesic activity and respiratory effects of butorphanol in sheep, *Research in Veterinary Science* 51, 19–23, 1991.

142. Schultz, R. A., Pretorius, P. J., and Terblanche, M., An electrocardiographic study of normal sheep using a modified technique, *Onderstepoort Journal of Veterinary Research* 39, 97–106, 1972.

143. Carroll, G. L. and Hartsfield, S. M., General anesthetic techniques in ruminants, *Veterinary Clinics of North America (Food Animal Practice)* 12, 627–661, 1996.

144. McCurnin, D. M. and Jones, R. L., Principles of surgical asepsis, in *Textbook of Small Animal Surgery*, Slatter, D., Ed., W. B. Saunders, Philadelphia, PA, 1993.

145. Andrews, E. J., Bennett, B. T., Clark, J. D., Houpt, K. A., Pascoe, P. J., Robinson, G. W., and Boyce, J. R., Report on the AVMA Panel on Euthanasia, *Journal of the American Veterinary Medical Association* 202, 229–249, 1993.

146. McNeal, L. G., Sheep handling concepts and equipment, *SID Sheep Research Digest* 2, 1–5, 1985.

147. Mews, A. R. and Mowlem, A., Goats, in *Practical Animal Handling*, Anderson, R. S. and Edney, A. T. B., Eds., Pergamon Press, Oxford, UK, 1991, 51–55.

148. Holmes, R. J., Sheep, in *Practical Animal Handling*, Anderson, R. S. and Edney, A. T. B., Eds., Pergamon Press, Oxford, UK, 1991, 39–49.

149. Bruns, D. P., Olmstead, M. L., and Litsky, A. S., Technique and results for total hip replacement in sheep: an experimental model, *Veterinary and Comparative Orthopaedics and Traumatology* 9, 158–164, 1996.

150. Chodobski, A., Szmydynger-Chodobska, J., Cooper, E., and McKinley, M. J., Atrial natriuretic peptide does not alter cerebrospinal fluid formation in sheep, *American Journal of Physiology* 262, R860-4, 1992.

151. Moseley, G. and Jones, J. R., A technique for sampling total rumen contents in sheep, *Research in Veterinary Science* 27, 97–98, 1979.

152. Andrews, E. J. and Hughes, H. C., Thromboembolic sequelae to indwelling silastic cannulas in sheep arteries, *Journal of Biomedical Materials Research* 7, 137–144, 1973.

153. Anderson, P. H., Nielsen, M. O., and Fjeldborg, J., Long-term carotid access in the goat: observations on application of a totally implantable catheter system, *Acta Veterinaria Scandinavica* 36, 579–581, 1995.

154. Ferrigno, M., Bishop, B., Impastato, K., Brady, A. G., Titford, M. E., and Bastian, F. O., Transvenous phrenic nerve stimulation during prolonged mechanical ventilation in awake, nonparalyzed ewes, *Laboratory Animal Science* 45, 690–693, 1995.

155. Kapp, G. M., Friend, T. H., Knabe, D. A., Bushong, D. M., and Clay, D., Nutritional, physiologic, and behavioral effects of metabolism crates on lambs, *Contemporary Topics in Laboratory Animal Science* 36, 61–65, 1997.

156. Falconer, J., Owens, P. C., and Smith, R., Cannulation of the cisterna magna in sheep: a method for chronic studies of cerebrospinal fluid, *Australian Journal of Experimental Biology and Medical Science* 63, 157–62, 1985.

157. Prelusky, D. B. and Hartin, K. E., A technique for serial sampling of cerebrospinal fluid from conscious swine and sheep, *Laboratory Animal Science* 41, 481–485, 1991.

158. Peregrine, A. S. and Mamman, M., A simple method for repeated sampling of lumbar cerebrospinal fluid in goats, *Laboratory Animals* 28, 391–396, 1994.

159. Hecker, J. F., *Experimental Surgery on Small Ruminants*, Butterworths, London, UK, 1974.

160. Lamerigts, N., Aspenberg, P., Buma, P., Versleyen, D., and Sloof, T. J. J. H., The repeated sampling bone chamber: a new permanent titanium implant to study bone grafts in the goat, *Laboratory Animal Science* 47, 401–406, 1997.

161. Houdebine, L. M., Production of pharmaceutical proteins from transgenic animals, *Journal of Biotechnology* 34, 269–287, 1994.

162. Adrian, T. E., Production of antisera using peptide conjugates, *Methods in Molecular Biology* 73, 239–249, 1997.

163. McLaughlin, C. L., Rogan, G. J., Buonomo, F. C., Cole, W. J., Hartnell, G. F., Hudson, S., Kasser, T. R., Miller, M. A., Baile, C. A. Finishing lamb performance responses to bovine and porcine somatotropins administered by Alzet pumps. *Journal of Animal Science* 69, 4039–4048, 1991.

164. Tilbrook, A. J., Galloway, D. B., Williams, A. H., and Clarke, I. J. Treatment of young rams with an agonist of GnRH delays reproductive development, *Hormones and Behavior* 27, 5–28, 1993.

165. Goodwin, S. D., Comparison of body temperatures of goats, horses, and sheep measured with a tympanic infrared thermometer, an implantable microchip transponder, and a rectal thermometer, *Contemporary Topics in Laboratory Animal Science* 37, 51–55, 1998.

166. Allen, M., Houlton, J. E. F., and Rushton, N., Total knee replacement in sheep: a new experimental model, *Veterinary Surgery* 22, 247, 1993.

167. Thomson, L. A., Houlton, J. E. F., Allen, M. J., and Rushton, N., Will the cranial cruciate ligament-deficient caprine stifle joint develop degenerative joint disease? *Veterinary and Comparative Orthopaedics and Traumatology* 7, 14–17, 1994.

168. Schreck, S., Inderbitzen, R., Chin, H., Wieting, D. W., Smilor, M., Breznock, E., and Pendray, D., Dynamics of Bjork-Shiley Convexo-Concave mitral valves in sheep, *Journal of Heart Valve Disease,* 4 Suppl. 1, S21–S24, 1995.

169. Wakisaka, Y., Taenaka, Y., Chikanari, K., Nakatani, T., Tatsumi, E., Masuzawa, T., Nishimura, T., Takewa, Y., Ohno, T., and Takano, H. Long-term evaluation of a non-pulsatile mechanical circulatory support system, *Artificial Organs* 21, 639–644, 1997.

170. Ofenloch, J. C., Chen, C., Hughes, J. D., and Lumsden, A. B., Endoscopic venous valve transplantation with a valve-stent device, *Annals of Vascular Surgery* 11, 62–67, 1997.

171. Schurmann, K., Vorwerk, D., Bucker, A., Neuerburg, J., Klosterhalfen, B., Muller, G., Uppenkamp, R., and Gunther, R. W., Perigraft inflammation due to Dacron-covered stent grafts in sheep iliac arteries: correlation of MR imaging and histopathologic findings, *Radiology* 204, 757–763, 1997.

172. Guillonneau, B., Wetzel, O., Lepage, J. Y., Vallencien, G., and Buzelin, J. M., Retroperitoneal laparoscopic nephrectomy: animal and human anatomic studies, *Journal of Endourology* 9, 487–490, 1995.

173. Dreyfus, M., Becmeur, F., Schwaab, C., Baldauf, J. J., Philippe, L., and Ritter, J., The pregnant ewe: an animal model for fetoscopic surgery, *European Journal of Obstetrics, Gynecology, and Reproductive Biology* 71, 91–94, 1997.

174. Newton, P. O., Cardelia, J. M., Farnsworth, C. L., Baker, K. J., and Bronson, D. G., A biomechanical comparison of open and thoracoscopic anterior spinal release in a goat model, *Spine* 23, 530–535, 1998.

175. Misiak, P. M. and Miceli, J. N., toxic effects of formalde-
     hyde, *Laboratory Management* 24, 63, 1968.

176. Formaldehyde Panel: Report of the Federal Panel on Form-
     aldehyde, National Toxicology Program, Research Triangle
     Park, 1980.

177. Johnson, D. D. and Libal, M. C., Necropsy of sheep and
     goats, *Veterinary Clinics of North America (Food Animal
     Practice)* 2, 129–146, 1986.

index

notes

T - #0583 - 101024 - C0 - 234/156/9 - PB - 9780849325687 - Gloss Lamination